Self-Regularity: A New Paradigm for Primal-Dual Interior-Point Algorithms

Self-Regularity: A New Paradigm for Primal-Dual Interior-Point Algorithms

Jiming Peng

Cornelis Roos

and

Tamás Terlaky

PRINCETON UNIVERSITY PRESS

PRINCETON AND OXFORD

Copyright © 2002 by Princeton University Press

Published by Princeton University Press,
41 William Street, Princeton, New Jersey 08540

In the United Kingdom: Princeton University Press,
3 Market Place, Woodstock, Oxfordshire OX20 1SY

Library of Congress Cataloging-in-Publication Data applied for
Peng, Jiming, Roos, Cornelis and Terlaky, Tamás
Self-Regularity: A New Paradigm for Primal-Dual Interior-Point Algorithms /
Jiming Peng, Cornelis Roos and Tamás Terlaky
p. cm.
Includes bibliographical references and index
ISBN 0-691-09193-5 (Paper)
0-691-09192-7 (Cloth)

British Library Cataloguing-in-Publication Data
A catalogue record for this book is available from the British Library

This book has been composed in Times and Abadi

Printed on acid-free paper

www.pup.princeton.edu

Printed in the United States of America

Contents

Preface

The primary goal of this monograph is to introduce a new framework for the theory of primal-dual interior-point methods (IPMs) based on the notion of the self-regularity of a function. Starting from Karmarkar's epoch-making paper [52] in 1984, the research on IPMs has dominated the field of continuous optimization for more than 15 years, and over 3000 papers have been published relating to IPMs. With the efforts of many excellent experts in the field, particularly the path-breaking work by Nesterov and Nemirovskii [83], the theory of IPMs has been developed into a rather mature principle.

An important concept in the IPM literature is the central path of the underlying problem. This was first recognized by Sonnevend [102] and Megiddo [70]. The primal-dual framework presented in [70] laid down the bedrock of many practically efficient IPMs with interesting theoretical properties. These IPMs utilize various strategies to follow the central path approximately. Two contrasting approaches in the analysis and implementation of IPMs are the so-called small-update (with small neighborhood) and large-update (with large neighborhood) methods. Unfortunately, there is still a big gap between the theory and the practical performance of IPMs with respect to these two strategies. As stated by Renegar [97, pp. 51], "It is one of the ironies of the IPM literature that algorithms which are more efficient in practice often have somewhat worse complexity bounds."

The motivation for this work is to bridge this gap. We deal with linear optimization, nonlinear complementarity problems, semidefinite optimization and second-order conic optimization problems. Our framework also covers large classes of linear complementarity problems and convex optimization. The algorithm considered may be interpreted as a path-following method or a constrained potential reduction method. Starting from a primal-dual strictly feasible point, our algorithm chooses a search direction defined by some Newton-type system derived from the self-regular proximity. The iterate is then updated, with the iterates staying in a certain neighborhood of the central path until an approximate solution to the problem is found. By exploring extensively some intriguing properties of the self-regular function, we establish that the complexity of large-update IPMs can come arbitrarily close to the best known iteration bounds of IPMs.

Acknowledgments

In the preparation of this monograph we greatly appreciated the support and stimulating discussions with many colleagues from all over the world.

First, we are indebted to Akiko Yoshise (University of Tsukuba, Japan) who made invaluable contributions to chapter 4 of this work. Our joint work resulted in some joint publications in this area.

Special thanks are due to those colleagues who helped us to clean the manuscript and to improve the presentation. Michael Saunders (Stanford University, USA), as always, made an extremely careful reading of an earlier version of the manuscript and provided us with countless corrections and suggestions for improvement. János Mayer (University of Zürich, Switzerland) gave numerous remarks and suggestions after a critical reading of large parts of the first draft of the monograph. Florian Jarre (University of Düsseldorf), Francois Glineur (Faculté Polytechnique de Mous, Belgium) and Florian Potra (University of Maryland) also gave us invaluable comments to improve the monograph. We kindly acknowledge the stimulating and encouraging discussions with Yurii Nesterov (CORE, C.U. Louvain-la-Neuve), Arkadii Nemirovskii (TECHNION, Haifa), Erling Andersen (EKA Consulting, Copenhagen) about various aspects of the self-regular paradigm. Their generous help is acknowledged here. In spite of our efforts and the invaluable support of our colleagues some typos might still remain in the monograph. The authors are solely responsible for all remaining errors.

We are also indebted to David Ireland of Princeton University Press for his continuing support while preparing the manuscript.

Developing a novel theory and writing a research monograph is a very time-consuming task: we want to thank our universities: McMaster University, Hamilton, Ontario, Canada and the Delft University of Technology, The Netherlands for giving us the opportunity to do this job.

The research of the first two authors was mainly supported by the project *High Performance Methods for Mathematical Optimization*, under the Dutch SWON grant 613-304-200. They are grateful for the support of the Faculty of Information Technology and Systems at Delft University of Technology, The Netherlands. Most of this work was done when the first author was visiting the third author in the Department of Computing and Software, McMaster

University, Canada. Support from that department is also acknowledged. Both the first and third authors were partially supported by the National Science and Engineering Research Council of Canada, grant # OPG 0048923 and by an FPP grant from IBM's T.J. Watson Research Laboratory.

Last but not least, we want to express warm thanks to our wives and children. They also contributed substantially to this monograph by their mental support, and by forgiving our shortcomings at home.

Jiming Peng
McMaster University

Cornelis Roos
Delft University of Technology

Tamás Terlaky
McMaster University

Notation

Notation for LO

A^{T}	transpose of the matrix A
C^k	the set of k-times differentiable functions
e	all-ones vector
E	the identity matrix in suitable space
I	index set $= \{1, 2, ..., n\}$
L	the input length of the LO problem
μ	the duality gap parameter
μ_0	initial duality gap
$\mathfrak{R}^n$	n-dimensional Euclidian vector space
$\mathfrak{R}^n_+ (\mathfrak{R}^n_{++})$	nonnegative (positive) orthant in $\mathfrak{R}^n$
x_i	ith coordinate of the vector x
x^{T}	transpose of x
$\|x\|$	the 2-norm of x
$\|x\|_p$	the p-norm of x
$x_{\max}$	the maximal component of x
$x_{\min}$	the minimal component of x
Δx	the primal part of the search direction
Δs	the dual part of the search direction
v	the scaled vector defined by $v_i = \sqrt{x_i s_i}/\sqrt{\mu}, \forall i \in I$
d_x	the primal part of the search direction in the scaled v-space: $d_x = v \Delta x/x$
d_s	the dual part of the search direction in the scaled v-space: $d_s = v \Delta s/s$
α	step size
ε	accuracy parameter
τ	proximity parameter
σ	the norm of the gradient of the proximity
θ	updating factor for μ
$\psi(t)$	a univariate function defining the proximity
$\psi(x)$	$:= (\psi(x_1), ..., \psi(x_n))^{\mathrm{T}}$, a mapping from $\mathfrak{R}^n$ into itself
$\Psi(x)$	the summation $\Psi(x) = \sum_{i=1}^{n} \psi(x_i)$

$\mathcal{N}(\mu, \tau)$ $:=\{(x, s) : \Psi(x, s, \mu) = \Psi(v) \leq \tau\}$, a neighborhood around the μ center

$\Upsilon_{p,q}(t)$ a sample self-regular proximity function

$$\Upsilon_{p,q}(t) = \frac{t^{p+1} - 1}{p(p + 1)} + \frac{t^{1-q} - 1}{q(q - 1)} + \frac{p - 1}{pq}(t - 1)$$

with $p, q \geq 1$

Ω_1 the set of functions that satisfy condition SR1 (page 31)

Ω_2 the set of functions that satisfy condition SR2 (page 31)

Notation for SOCO

K $= \{x : x \in \mathfrak{R}^n, x_1 \geq \|(x_2, x_3, ..., x_n)\|\}$

K_+ $= \{x : x \in \mathfrak{R}^n, x_1 > \|(x_2, x_3, ..., x_n)\|\}$

$\tilde{e}$ $= (1, 0, ..., 0)^T$

J $:= \{1, 2, ..., N\}$ where N is the number of the constrained cones

$x \succ_K 0 \ (x \succeq_K 0)$ $x \in K_+$ (or $x \in K$)

$x_{2:n}$ $(x_2, x_3, ..., x_n)^T$

$\lambda_{\max}(x)$ $:= x_1 + \|x_{2:n}\|$

$\lambda_{\min}(x)$ $:= x_1 - \|x_{2:n}\|$

$\lambda_{\max}(|x|)$ $:= |x_1| + \|x_{2:n}\|$

$\lambda_{\min}(|x|)$ $:= |x_1| - \|x_{2:n}\|$

$\text{mat}(x)$ a matrix generated by the components of x (page 126)

$\psi(x)$ a mapping associated with the second-order cone K (page 132)

$\det(x)$ $:= \lambda_{\max}(x)\lambda_{\min}(x)$, the determinant of x (page 131)

$\text{Tr}(x)$ $:= \lambda_{\max}(x) + \lambda_{\min}(x) = 2x_1$, trace of x (page 131)

$\Psi(x)$ $:= \text{Tr}(\psi(x))$, trace of $\psi(x)$

$x(t)$ $= (x_1(t), x_2(t), ..., x_n(t))^T$

$x'(t)$ $= (x_1'(t), x_2'(t), ..., x_n'(t))^T$

$\Delta\psi(\cdot, \cdot)$ the divided difference

$$\Delta\psi(t_1, t_2) = \frac{\psi(t_1) - \psi(t_2)}{t_1 - t_2} \quad \text{if } t_1 \neq t_2 \text{ and } \Delta\psi(t, t) = \psi'(t)$$

Notation for SDO

$\mathfrak{R}^{n \times n}$ space of $n \times n$ real matrices

$S^{n \times n}$ space of symmetric $n \times n$ matrices

$S_+^{n \times n} \ (S_{++}^{n \times n})$ space of symmetric positive semidefinite (definite) $n \times n$ matrices

$A \succ 0 \; (A \succeq 0)$	$A \in S_{++}^{n \times n} \; (A \in S_{+}^{n \times n})$
$A \prec 0 \; (A \preceq 0)$	$-A \in S_{++}^{n \times n} \; (-A \in S_{+}^{n \times n})$
$\lambda_i(A)$	eigenvalues of $A \in S^{n \times n}$ ordered according to the decreasing values
$\varrho_i(A)$	singular values of A ordered according to the decreasing values
$\lambda_{\max}(A)$	the maximal eigenvalue of A
$\lambda_{\min}(A)$	the minimal eigenvalue of A
A_{ij}	ijth entry of A
$\mathrm{Tr}(A)$	the trace of A, $\mathrm{Tr}(A) = \sum_{i=1}^{n} A_{ii}$
$\|A\|$	Frobenius norm of A
$\psi(X)$	a function from $S^{n \times n}$ into itself (page 103)
$\Psi(X)$	$:= \mathrm{Tr}(\psi(X))$, trace of $\psi(X)$
$X(t)$	a matrix-valued function of t with $X(t) = [X_{ij}(t)]$
$X'(t)$	the gradient of $X(t)$ with $X'(t) = [X'_{ij}(t)]$

List of Abbreviations

CP(s)	Complementarity Problem(s)
IPC	Interior-Point Condition
IPM(s)	Interior-Point Method(s)
LCP(s)	Linear Complementarity Problem(s)
LO	Linear Optimization
NCP(s)	Nonlinear Complementarity Problem(s)
QO	Quadratic Optimization
SDO	Semidefinite Optimization
SOCO	Second-Order Conic Optimization
VIP(s)	Variational Inequality Problem(s)

Self-Regularity: A New Paradigm for Primal-Dual Interior-Point Algorithms

Chapter 1

Introduction and Preliminaries

Linear Programming is a revolutionary development that permits us, for the first time in our long evolutionary history, to make decisions about the complex world in which we live that can approximate, in some sense, the optimal or best decision.

George B. Dantzig [80]

A short survey about the fields of linear optimization[1] and interior-point methods is presented in this chapter. Based on the simple model of standard linear optimization problems, some basic concepts of interior-point methods and various strategies used in the algorithm are introduced. The purpose of this work, as well as some intuitive observations that sparked the authors' research, are described. Several preliminary technical results are presented as a preparation for later analysis, and the contents of the book are also outlined.

---

[1] For some historical reason, the name "linear programming" was coined in the early stages of the field of linear optimization. In this monograph we choose its more natural name "linear optimization". This also distinguishes the field from general programming works related to computers.

1.1 HISTORICAL BACKGROUND OF INTERIOR-POINT METHODS

1.1.1 Prelude

There is no doubt that the major breakthroughs in the field of mathematical programming are always inaugurated in linear optimization. Linear optimization, hereafter LO, deals with a simple mathematical model that exhibits a wonderful combination of two contrasting aspects: it can be considered as both a continuous and a combinatorial problem. The continuity of the problem is finding a global minimizer of a continuous linear function over a continuous convex polyhedral constrained set, and its combinatorial character is looking for optimality over a set of vertices of a polyhedron. The Simplex algorithm [17], invented by Dantzig in the mid-1940s, explicitly explores its combinatorial structure to identify the solution by moving from one vertex to an adjacent one of the feasible set with improving values of the objective function. Dantzig's Simplex method has achieved tremendous success in both theory and application, and it remains to this day one of the efficient workhorses for solving LO.

The invention of the digital computer has brought dramatic changes to the world, and thus to the area of optimization. By combining computers with efficient methods, we can now solve many large and complex optimization problems that were unsolvable before the advent of computers, or even two decades ago. Inspired by the fast development of computers, scientists began to study the complexity theory of methods in the 1960s and 1970s. An algorithm was termed polynomial if the number of arithmetic operations taken by the algorithm to solve the problem, could be bounded above by a polynomial in the input "size" of the problem.[2] Correspondingly such a polynomial algorithm was recognized as an efficient algorithm. Although the Simplex method with certain pivot rules works very well in practice and its probabilistic computational complexity is strongly polynomial (see Borgwardt [12]), it might take $2^n - 1$ iterations to solve the LO problem constructed by Klee and Minty [54] in 1972, where n is the number of variables in the problem with $2n$ constraints. Similar "exponential examples" for different variants of the Simplex method have also been reported. Agitated by these difficult exponential examples for the Simplex method, researchers in the field became interested in the issue of whether LO problems are solvable in polynomial time.

---

[2] The input size of a problem (denoted by L) indicates the length of a binary coding of the input data for the problem. For the standard LO (1.1) on page 5, the length is $L = \sum_{i=0}^{m} \sum_{j=0}^{n} \lceil \log_2(|a_{ij}| + 1) \rceil$ with $a_{i0} = b_i$, $a_{0j} = c_j$ and $a_{00} = 0$. In this work, we derive upper bounds for the number of iterations to obtain an ε-optimal solution of the problem under consideration, where ε is an accuracy parameter used in the algorithm for termination.

In 1979, an affirmative answer to the above question was given by Khachiyan [53], who utilized the so-called ellipsoid method to solve the LO problem and proved that the algorithm has a polynomial $\mathcal{O}(n^2L)$ iteration complexity with a total of $\mathcal{O}(n^4L)$ bit operations. Khachiyan's results were immediately hailed in the international press, and the ellipsoid algorithm was subsequently studied intensively by various scholars in both theory and implementation. Unfortunately, in contrast to people's high expectation, even the best known implementation of the ellipsoid method is far from competitive with existing Simplex solvers. In this first challenge of polynomial algorithms to the Simplex method, the latter was the obvious winner. Nevertheless, the sharp contrast between the practical efficiency and the theoretical worst-case complexity of an algorithm set the stage for further exciting developments.

1.1.2 A Brief Review of Modern Interior-Point Methods

The new era of interior-point methods (IPMs) started in 1984 when Karmarkar [52] proposed his LO algorithm, which enjoyed a polynomial complexity of $\mathcal{O}(nL)$ iterations with $\mathcal{O}(n^{3.5}L)$ bit operations, and made the announcement that his algorithm could solve large-scale LO problems much faster than the Simplex method. At that time, the name of Karmarkar and his algorithm reached the front page of the *New York Times* despite the fact that his claim was received with much skepticism by some experts in the field. Nowadays, it is clear that Karmarkar opened a new field: the flourishing field of modern IPMs.

Karmarkar's algorithm in its original form is a primal projective potential reduction method. The potential function was employed to measure the progress of the algorithm and to force the iterates to stay in the feasible region. Although Karmarkar considered a nonstandard LO model, it soon turned out that the projective transformation in [52] is not necessary for solving standard LO problems. Shortly after the publication of [52], Gill et al. [28] observed that some simple variants of Karmarkar's algorithm could be tracked back to a very old algorithm in nonlinear optimization: *the logarithmic barrier method*. This observation led to a revival of some old methods for continuous optimization including *the logarithmic barrier method* by Frisch [24,25] and Fiacco and McCormick [23], *the center method* by Huard [41], and *the affine scaling method* by Dikin [18]. For example, it was proven by Roos and Vial [99] that the basic logarithmic barrier method for LO has a polynomial complexity. Most of the early work on IPMs in the 1980s followed Karmarkar's primal setting, but focused on more efficient implementation or better complexity bounds. A remarkable work at that time was the algorithm [96] by Renegar, who proposed using upper bounds on the optimal value of the objective function to form successively smaller subsets of the feasible set, and employ Newton's method to follow the analytic centers of these subsets to get the primal optimal solution. Closely related to the center of a polyhedron is another very important concept in the IPM literature: the so-called *central path* first recognized by Sonnevend [102] and Megiddo [70]. Almost all known polynomial-time variants of IPMs use the central path as a guideline to the optimal set, and some variant of Newton's method to follow the central path approximately. These Newton-type methods fall into different groups with respect to the strategies used in the algorithms to follow the central path. The reader may consult the survey paper by Gonzaga [33] or the monograph of den Hertog [39] for further details.

Among versatile types of IPMs, the so-called primal-dual path-following methods have emerged as particularly attractive and useful. The primal-dual setting was first suggested by Megiddo [70], Monteiro and Adler [75], Tanabe [106] and Kojima et al. [59] for LO problems, and later extensively investigated for complementarity problems by a Japanese group led by Kojima [58]. In the theoretical aspect, primal-dual IPMs possess the appealing theoretical polynomiality and allow transparent extension to other problems such as convex programming and complementarity problems. In practice they also form the basis of most efficient IPM solvers. A notable work with respect to the practical IPM is due to Mehrotra [71] who described a predictor-corrector algorithm in 1992. Mehrotra's scheme has been proved to be very effective in practice and is employed in most existing successful IPM solvers. Much effort went into investigating preprocessing techniques, warm start, sparse Cholesky factorization, and other implementation issues. By these efforts, the implementation of IPMs was enhanced greatly at both the commercial and the academic level. The paper [7] present a thorough survey of the implementation and performance of IPMs. As claimed there, compared with the Simplex method, IPMs appear to be a strong rival for solving LO problems of medium size, and the winner for large-scale ones (see also [66]).

The theory of IPMs developed into a mature principle during the 1990s. One main contribution to IPM theory in that period came from two mathematicians, Nesterov and Nemirovskii [83], who invented the theory of the, by now, well-known self-concordant functions, allowing the algorithms based on the logarithmic barrier function for LO to be transparently extended to more complex problems such as nonlinear convex programming, nonlinear complementarity problems, variational inequalities, and particularly to semidefinite optimization (SDO) and second-order conic optimization (SOCO). Nesterov and Todd [84,85] further generalized the primal-dual algorithms to linear optimization over so-called self-scaled cones, which still include SDO and SOCO as concrete instances. To some extent, SDO recently became the most active area of mathematical programming. SDO involves minimizing the values of a linear function with a matrix argument subject to linear equality constraints and requiring that the matrix argument be positive semidefinite. The SDO paradigm includes as special cases LO, quadratic optimization

(QO), the linear complementarity problem (LCP), and SOCO. It has a great variety of applications in various areas such as optimal control, combinatorics, structural optimization, pattern recognition, etc. We refer to the excellent survey by Vandenberghe and Boyd [118] for more details. An extremely important fact is that, while problems such as LO, QP and LCP can also be solved by other methods (and hence IPM is only one alternative), IPMs appear to be the first and also most efficient approach for SDO. As such, SDO is a good advertisement for the power of IPMs, although the theory and especially implementation of IPMs for SDO are still far from mature. There are also many studies of other theoretical properties of IPMs such as local convergence of the algorithm, procedures to get an exact solution from an approximate solution and sensitivity analysis. Interested readers are referred to recent books by Roos, Terlaky and Vial [98], Wright [122] and Ye [127] and the references therein. The book edited by Terlaky [107] collects some state-of-the-art review papers on various topics of IPMs and a large number of related references. For more recent achievements on IPMs we refer to the Interior-Point Methods Online Web site, *www.mcs.anl.gov/otc/InteriorPoint*, where most of the technical reports in the past five years are listed.

1.2 PRIMAL-DUAL PATH-FOLLOWING ALGORITHM FOR LO

1.2.1 Primal-Dual Model for LO, Duality Theory and the Central Path

In this monograph we are mainly concerned with the complexity theory of IPMs. This is the major theme in the recent IPM literature. We start with the following standard linear optimization problem:

$$\text{(LP)} \quad \min\{c^{\mathsf{T}}x : Ax = b, x \geq 0\}, \tag{1.1}$$

where $A \in \Re^{m \times n}$, $b \in \Re^m$, $c \in \Re^n$, and its dual problem

$$\text{(LD)} \quad \max\{b^{\mathsf{T}}y : A^{\mathsf{T}}y + s = c, s \geq 0\}. \tag{1.2}$$

Throughout this work, we assume that A has full row rank, that is, rank$(A) = m$. This implies that for a given dual feasible s, the vector y is uniquely defined. Hence we may identify a feasible solution of (LD) only by s. The following definitions introduce some basic concepts for LO.

Definition 1.2.1 *A point x is said to be feasible (or strictly feasible) for (LP) if $Ax = b$ and $x \geq 0$ (or $x > 0$). A point (y, s) is feasible (or strictly feasible) for (LD) if $A^{\mathsf{T}}y + s = c$ and $s \geq 0$ (or $s > 0$). The point (x, y, s) is primal-dual feasible (or strictly feasible) if x and (y, s) are feasible (or strictly feasible) for (LP) and (LD), respectively.*

The relationships between (LP) and (LD) have been well explained by the duality theory of LO. For instance, if (x, y, s) is a primal-dual feasible pair, then there holds $b^{\mathrm{T}}y \leq c^{\mathrm{T}}x$. In other words, the objective value in (LD) gives a lower bound for the objective in (LP), and the objective in (LP) provides an upper bound for that in (LD). The main duality results can be summarized by the following strong duality theorem [45,98].

Theorem 1.2.2 *For (LP) and (LD) one of the following four alternatives holds:*

(i) *(LP) and (LD) are feasible and there exists a primal-dual feasible pair (x^*, y^*, s^*) such that*

$$c^{\mathrm{T}}x^* = b^{\mathrm{T}}y^*.$$

(ii) *(LP) is infeasible and (LD) is unbounded.*

(iii) *(LD) is infeasible and (LP) is unbounded.*

(iv) *Both (LP) and (LD) are infeasible.*

Hence, solving LO amounts to detecting which of these four cases holds, and in case (i) an optimal solution (x^*, y^*, s^*) must be found. Note that in case (i), the two objective values in (LP) and (LD) coincide with each other at the solution (x^*, y^*, s^*), that is $c^{\mathrm{T}}x^* = b^{\mathrm{T}}y^*$, which further yields

$$(s^*)^{\mathrm{T}}x^* = (c - A^{\mathrm{T}}y^*)^{\mathrm{T}}x^* = c^{\mathrm{T}}x^* - b^{\mathrm{T}}y^* = 0.$$

Observe that since $x^*, s^* \geq 0$, the above equality can also be written as

$$x_i^* s_i^* = 0, \quad i = 1, \ldots, n.$$

An intrinsic property of LO is given by the following result.

Theorem 1.2.3 *Suppose that both (LP) and (LD) are feasible. Then there exists a primal-dual feasible pair (x^*, y^*, s^*) such that $(x^*)^{\mathrm{T}}s^* = 0$ and $x^* + s^* > 0$. A solution (x^*, s^*) with this property is called strictly complementary.*

This theorem was first established by Goldman and Tucker [32] and later studied by other researchers using different approaches. It plays an important role in the design and analysis of IPMs, particularly in the procedures for getting an exact solution from an approximate solution obtained by IPMs, or detecting the infeasiblity of the problem [98,127].

Starting from an initial point (x^0, y^0, s^0) with $x^0, s^0 > 0$, all primal-dual interior-point algorithms generate a point sequence (x^k, y^k, s^k) with $x^k, s^k > 0$ such that the sequence converges to the set of optimal solutions as k goes to infinity. If, at each iteration, the point (x^k, y^k, s^k) further satisfies the linear equality constraints, then we call the algorithm a *feasible interior-point algorithm*, and otherwise an *infeasible interior-point algorithm*. The choice between feasible and infeasible IPMs depends on whether a feasible starting point is available or not. If a strictly feasible point is known, then by starting from this point and carefully updating the iterates, we can keep the iterative sequence feasible.

In this work, the class of feasible IPMs is taken as the framework for our discussions. The reasons for this are twofold. First, from a theoretical viewpoint it is more convenient to analyze a feasible IPM rather than an infeasible one. Needless to say, with some extra effort, the analysis for feasible IPMs can usually be extended to its infeasible counterpart. Second, for some problems such as LO, LCP, and SDO, by slightly increasing the size of the problem, we can always use the self-dual embedding model (see Chapter 7 or [55,98,127]) to reformulate the original problem as a new problem for which a strictly feasible starting point is readily available. Hence from now on we assume without loss of generality that both (LP) and (LD) satisfy *the interior-point condition*, that is, there exists $(x^0, y^0, s^0,)$ such that

$$Ax^0 = b, \quad x^0 > 0, \quad A^\mathsf{T}y^0 + s^0 = c, \quad s^0 > 0. \tag{1.3}$$

By using the self-dual embedding model (Chapter 7), we can further assume that $x^0 = s^0 = e$. Under the interior-point condition, we are in case (*i*) of Theorem 1.2.2, and hence an optimal solution pair always exists. From Theorem 1.2.2 one can see that finding an optimal solution of (LP) and (LD) is equivalent to solving the following system:

$$Ax = b, \quad x \geq 0,$$

$$A^\mathsf{T}y + s = c, \quad s \geq 0, \tag{1.4}$$

$$xs = 0.$$

Here xs denotes the coordinatewise product of the vectors x and s. The basic idea of primal-dual IPMs is to replace the third equation in (1.4), the so-called *complementarity condition* for (LP) and (LD) by the parameterized equation $xs = \mu e$, where e denotes the all-one vector and $\mu > 0$. Thus we consider the system

$$Ax = b, \quad x \geq 0,$$

$$A^\mathsf{T}y + s = c, \quad s \geq 0, \tag{1.5}$$

$$xs = \mu e.$$

The existence of a unique solution to the above system is well-known (see McLinden [68], Kojima et al. [59] and Güler [35]).

Theorem 1.2.4   *If the interior-point condition (1.3) holds, then for each* $\mu > 0$, *the parameterized system (1.5) has a unique solution.*

Let us denote this solution by $(x(\mu), y(\mu), s(\mu))$. We call $x(\mu)$ the μ-center of (LP) and $(y(\mu), s(\mu))$ the μ-center of (LD). The set of μ-centers (with μ running through all positive real numbers) gives a homotopy path, which is called the central path of (LP) and (LD) (Sonnevend [102], Megiddo [70]). The limiting behavior of the central path as μ goes to zero has been a hot topic for some time. In [68], McLinden investigated the limiting behavior of the path for monotone complementarity problems. The properties of the central path for LO were first considered by Megiddo [70], and later by Güler and Ye [36]. In the following, we discuss some of the main properties of the central path. To this end, we first introduce the definition of the analytic center of the optimal set for LO.

Definition 1.2.5   *Let*

$$\mathcal{F} = \{(x, s) : Ax = b,\ A^{\mathrm{T}}y + s = c,\ x \geq 0,\ s \geq 0,\ x^{\mathrm{T}}s = 0\}$$

denote the optimal solution set of primal-dual pair (LP) and (LD). Let

$$\mathcal{B}_{\mathcal{F}} = \{i : \exists (x, s) \in \mathcal{F} \text{ such that } x_i > 0\},$$

$$\mathcal{N}_{\mathcal{F}} = \{i : \exists (x, s) \in \mathcal{F} \text{ such that } s_i > 0\}.$$

The analytic center of $\mathcal{F}$ *is given by*

$$(x^*, s^*) = \arg\max_{(x,s) \in \mathcal{F}} \prod_{i \in \mathcal{B}_{\mathcal{F}}} x_i \prod_{j \in \mathcal{N}_{\mathcal{F}}} s_j.$$

It follows from Theorem 1.2.3 that the analytic center of $\mathcal{F}$ is a strictly complementary solution for LO. A nice property of the central path is that it converges not only to the optimal set $\mathcal{F}$, but also to its analytic center [98].

Theorem 1.2.6   *The central path converges to the analytic center of the optimal set* $\mathcal{F}$ *as* $\mu \to 0$.

1.2.2 Primal-Dual Newton Method for LO

All the primal-dual path-following algorithms trace the central path approximately. Let us briefly indicate this. To outline a general procedure for IPMs, we first need to define a so-called *neighborhood of the* μ-center:

$$\mathcal{N}(\tau, \mu) = \{(x, s) > 0 : Ax = b,\ A^{\mathrm{T}}y + s = c,\ \eta(x, s, \mu) \leq \tau\},$$

where $\eta(x, s, \mu)$ is a so-called *proximity* to measure the distance from the point (x, s) to $(x(\mu), s(\mu))$, and τ is the *radius of the neighborhood*. For the time being, we focus on the structure of the algorithm and leave the discussion of various choices of the proximity $\eta(x, s, \mu)$ to the next section. Most feasible IPMs take the following form:

General Feasible Primal-Dual Newton Method for LO

Input:
  A proximity parameter τ,
  an accuracy parameter $\varepsilon > 0$;
  a fixed barrier update parameter θ, $0 \le \theta \le 1$;
  (x^0, s^0) and $\mu^0 = 1$ such that $\eta(x^0, s^0, \mu^0) \le \tau$.
begin
  $x := x^0$; $s := s^0$; $\mu := \mu^0$;
  while $n\mu \ge \varepsilon$ **do**
  begin
    if $\eta(x, s, \mu) \le \tau$ **then**
      begin (Outer Iteration)
        $\mu := (1 - \theta)\mu$;
      end
    end if
    begin (Inner Iteration)
      Solve system (1.6) for $\Delta x, \Delta y, \Delta s$;
      Determine step size α by some rules;
      $x := x + \alpha\Delta x$;
      $y := y + \alpha\Delta y$;
      $s := s + \alpha\Delta s$;
      Update μ by certain rules.
    end
  end
end

Remark 1.2.7 To be consistent with the name used in [98], we call the step where the present iterate is in a certain neighborhood of the current μ-center an *outer iteration* and the procedure to get a primal-dual pair (x, y, s) in the neighborhood of this μ-center an *inner iteration*. In the algorithm, we use the proximity $\eta(x, s, \mu)$ to control the iterates. Various choices of these proximities are discussed in the next section.

Without loss of generality we assume that a point (x, y, s) in the neighborhood $\mathcal{N}(\tau, \mu)$ is known for some positive μ.[3] Hence, in each outer iteration we need to update μ by $\mu := (1 - \theta)\mu$ first and then solve the following Newton system:

$$A\Delta x = 0,$$

$$A^{\mathrm{T}}\Delta y + \Delta s = 0, \tag{1.6}$$

$$s\Delta x + x\Delta s = \mu e - xs.$$

Because A has full row rank, system (1.6) has a unique solution for any $(x, s) > 0$. Correspondingly we get a search direction $(\Delta x, \Delta y, \Delta s)$ by solving system (1.6).

In some IPMs, the parameter θ is not fixed but chosen as a number in $[0, 1]$ corresponding to the present iterate. In the case $\theta = 1$, the solution in the outer iteration is called the *affine scaling direction* or *predictor direction*. If $\theta = 0$, then we obtain the so-called *corrector direction* or *centering direction*. By taking a step along the search direction, where the step size α is defined by some line search rules, one constructs a new triple $(x, y, s) = (x_+, y_+, s_+)$ with $x_+ = x + \alpha\Delta x$, $y_+ = y + \alpha\Delta y$ and $s_+ = s + \alpha\Delta s$. Then μ is updated by certain rules. In most IPMs, μ is updated corresponding to the present duality gap, that is, $\mu = x^{\mathrm{T}}s/n$. After the update of μ, then we want to check whether the current iterate is in the neighborhood $\mathcal{N}(\tau, \mu)$ of the new center $(x(\mu), y(\mu), s(\mu))$. If the answer is "yes", then go to the outer iteration and repeat the Newton process until the duality gap is sufficiently small. Otherwise, we need to run the inner process until the iterative point enters the neighborhood $\mathcal{N}(\tau, \mu)$ again. Then we go back to outer iteration, and so on. This process is repeated until an approximate solution to the problem is obtained. Most practical algorithms then construct an exact solution by resorting to a rounding procedure as described by Ye [124] (see also Mehrotra and Ye [72] or [98]), and possibly produce a basic solution via a basis identification procedure or by crossing over to the Simplex method.

For some IPMs, the step size is chosen very carefully so that the new iteration will always stay in a certain neighborhood of the central path. Hence for these algorithms, only one inner iteration is needed at each outer iteration. A more practical way to deal with the step size α is that, in each outer iteration, where usually a large θ is employed,[4] we compute the maximal feasible step size $(\alpha_{\max})$ first and then use $\alpha_{\max}$ up to a certain ratio (for instance $0.9\alpha_{\max}$) as a step size. This step aims to reduce the duality gap as much as possible while keeping the iterate strictly feasible. If it is necessary, we use the inner process to get a point in the neighborhood $\mathcal{N}(\tau, \mu)$ again. The idea of utilizing a fixed factor of the maximal step size as the working step

---

[3] By using the self-dual embedding model (see Chapter 7), we can get a starting point exactly on the central path of the embedded problem with $\mu = 1$.

[4] If $\theta = 1$ and μ is defined by $\mu = x^{T}s/n$ after each step, then we get the so-called predictor-corrector IPMs.

size has been widely employed in most IPM packages and works extremely well [7].

We proceed with a discussion of the inner process. Note that the only difference between an outer iteration and an inner iteration is that in the outer iteration we need to update μ first. In the inner process, if we update the parameter μ by $\mu = x^\mathrm{T} s/n$ after each step, then from (1.6) we get $\Delta x^\mathrm{T} \Delta s = 0$. Further, we can see that for any solution of (1.6), the following relation holds:

$$x^\mathrm{T} \Delta s + s^\mathrm{T} \Delta x = n\mu - x^\mathrm{T} s = 0 \quad \text{if } \mu = \frac{x^\mathrm{T} s}{n}.$$

This indicates that during the inner process, the new duality gap after one step satisfies $(x + \alpha \Delta x)^\mathrm{T}(s + \alpha \Delta s) = x^\mathrm{T} s$. Because μ acts indeed as an independent parameter in the inner process, the above iterative process can therefore be interpreted in the following alternative way: in each outer iteration we update the parameter μ by $\mu = (1 - \theta)\mu$ first and then solve system (1.6); and we do not update μ in the inner procedure. We can go a little farther. We assume that in both outer and inner iterations, the parameter μ is independent of the duality gap and we only update μ in the outer iterations. This model has been employed for most algorithms in the book by Roos, Terlaky and Vial [98]. We choose this model as the subject of the investigation in the present work. The reason for this is that such a model is very convenient for the complexity analysis of IPMs. In addition, the analysis for such a model where μ is cast as a free parameter can always be adapted to handle the model where μ is defined via the current iterate $\mu = x^\mathrm{T} s/n$ without much difficulty. Lastly we would like to emphasize again that the parameter μ is treated as an independent parameter in the inner iterations in the practical implementation of IPMs.

Since we cast μ as an independent parameter in the iteration, to avoid some unnecessary confusing arguments, we describe the classical primal-dual Newton method as follows:

Classical Primal-Dual Newton Method for LO

Inputs
    A proximity parameter τ;
    an accuracy parameter $\varepsilon > 0$;
    a fixed barrier update parameter θ, $0 \leq \theta \leq 1$;
    (x^0, s^0) and $\mu^0 = 1$ such that $\eta(x^0, s^0, \mu^0) \leq \tau$.
begin
    $x := x^0$; $s := s^0$; $\mu := \mu^0$;
    while $n\mu \geq \varepsilon$ **do**
    begin
        $\mu := (1 - \theta)\mu$;

```
while n(x, s, μ) ≥ τ do
begin
    Solve system (1.6) for Δx, Δy, Δs;
    Determine a step size α;
    x := x + αΔx;
    y := y + αΔy;
    s := s + αΔs;
end
end
end
```

1.2.3 Strategies in Path-following Algorithms and Motivation

It is clear from the description in the last section that to continue the iteration process, we want to know when a point (x, y, s) is 'close' enough to the current μ-center, so that we can update the parameter μ and go on to the next outer iteration. In the classical primal-dual Newton method, this is done by means of two arguments: the proximity $\eta(x, s, \mu)$ and the constant τ.

Two obvious choices for η in the IPM field are the Euclidean norm $\|(xs/\mu) - e\|$ and the infinity norm $\|(xs/\mu) - e\|_\infty$, while τ is always chosen as a small constant such as 1/2 [98,122,127]. In the IPM literature, the neighborhood defined by the infinity norm is usually referred to as a large neighborhood, and a small neighborhood if the Euclidean norm is applied.

The choice of the proximity measure is crucial not only for the quality and elegance of the analysis, but also for the performance of the algorithm. In practice, IPMs with large neighborhood are always more efficient than IPMs with small neighborhood. In sharp contrast, IPMs working with small neighborhood have better worst-case complexity iteration bounds than IPMs with large neighborhood.

Besides the two proximities mentioned above, several other proximities have been introduced and used in the IPM literature [98,122,127]. These proximities share some desirable properties that are very helpful in the analysis of IPMs. For simplicity, let us first introduce some notation. For any strictly feasible primal-dual pair (x, s) and any positive number μ, we define

$$v := \sqrt{\frac{xs}{\mu}}, \quad v^{-1} := \sqrt{\frac{\mu e}{xs}} \tag{1.7}$$

to be the vectors whose ith components are $\sqrt{x_i s_i/\mu}$ and $\sqrt{\mu/(x_i s_i)}$ respectively.[5] Note that by using this notation, we can write the centrality

---

[5] This notation is widely used in the IPM literature to ease the description and analysis of IPMs.

condition $xs = \mu e$ as $v^2 = e$ or equivalently $v = e$. Many proximity measures or potential functions used in the IPM literature can be expressed as a function of v.

The following are two popular proximity measures for primal-dual IPMs [98,122,127]:

$$\delta(x, s, \mu) := \frac{1}{\sqrt{2}} \| v - v^{-1} \|, \tag{1.8}$$

$$\Phi(x, s, \mu) := \sum_{i=1}^{n} \phi(v_i) = \frac{x^T s}{2\mu} - \frac{n}{2} + \frac{n}{2} \log \mu - \frac{1}{2} \sum_{i=1}^{n} \log(x_i s_i), \tag{1.9}$$

where

$$\phi(t) = \frac{1}{2} t^2 - \frac{1}{2} - \log t.$$

It is easy to see that both proximity measures are strictly positive for any $v \neq e \in R_{++}^n$ and vanishes only when $v = e$. In other words, the centrality condition $xs = \mu e$ represents precisely the optimality condition for these two proximities. Moreover, both proximities go to infinity if v approaches the boundary of the nonnegative orthant. The latter case is characterized as the *barrier property* of the proximity. The measure Φ is closely related to the *logarithmic barrier function* with the *barrier parameter* μ; its usefulness has been known for a long time (cf. Frisch [24], Lootsma [64] and Fiacco and McCormick [23]). With the exception of the relation between the centrality condition $xs = \mu e$ and the optimality condition for the function Φ, the search direction based on the standard Newton method can be interpreted as the steepest descent direction for the logarithmic barrier function in the scaled v-space. These fascinating features of the logarithmic barrier function illustrate that it is at the very heart of classical Newton-step based IPMs.

The measure δ, up to a factor $\sqrt{2}/2$, was introduced by Jansen et al. [47] and thoroughly used in [89,90,98,133]. Its SDO analogue was also used in the analysis of interior-point methods for semidefinite optimization [55]. We note that variants of the proximity $\delta(x, s, \mu)$ were used by Kojima et al. in [58] and Mizuno et al. in [73]. There are also many others proximities used in the IPM literature [98,122,127]. Usually these proximities are more or less related to a special class of functions: the so-called *self-concordant* barrier functions introduced by Nesterov and Nemirovskii [83].

Another important ingredient of IPMs is the choice of the parameter θ. Usually, if θ is a constant independent of n (the dimension of the problem), for instance $\theta = 1/2$, then we call the algorithm a large-update (or long-step) method. If θ depends on the problem dimension such as $\theta = 1/(2\sqrt{n})$, then the algorithm is named a small-update (or short-step) method. As we stated in the preface, at present there is still a gap between the practical performance and the theoretical worst-case complexity of these two classes of IPMs in the literature. This is especially true for primal-dual large-update methods, which are the most efficient methods in practice (see, e.g., Andersen et al. [7]). In what follows, we explain this phenomenon in more detail.

We begin by considering a specific small-update IPM with the parameters $\theta = 1/(2\sqrt{n})$ and $\tau = 1$. Assume that the starting point lies in the neighborhood of the central path $\mathcal{N}(\mu, \tau) = \{(x, s) : \delta(x, s, \mu) \leq \tau\}$. For simplicity, we denote by δ and δ_+ the proximity at the present iterate and after one update, respectively. Therefore we can start an inner iteration. It has been shown (see, for instance, Chapter 7 in [98]) that in this special situation, the full Newton step is feasible and after such a full Newton step, the proximity satisfies $\delta_+ \leq \delta^2/2 \leq 1/2$. Hence we can update μ and check whether the point is in the neighborhood of the new center. Note that, after the update of μ, the proximity δ might increase. However, it has been proven [98] that, if $\delta \leq 1/2$ and $\theta = 1/(2\sqrt{n})$, then the increased proximity still satisfies $\delta_+ \leq 1$, and hence it stays within $\mathcal{N}(\mu, \tau)$. Thus we can dispense with the inner iteration and move to the next outer iteration. Observe that in this process, we reduce the parameter μ by a fixed ratio. It follows immediately if k is a constant such that $n(1 - \theta)^k \mu_0 \leq \varepsilon$, then the algorithm will terminate after k iterations and report an ε-solution to the problem. By using some simple calculus, one can show that $k \leq 2\sqrt{n} \log(n\mu_0/\varepsilon)$ for sufficiently large n (Theorem II.52 in [98]) and this leads to the conclusion that the small-update method has an iteration bound $\mathcal{O}(\sqrt{n} \log(n/\varepsilon))$, which is to date the best known complexity result for IPMs.

Now let us indicate how a typical large-update method works, for instance $\theta = 0.9$ and $\tau = \sqrt{n}$. Suppose that the current point is in the neighborhood $\mathcal{N}(\mu, \tau)$. Then we can set $\mu = 0.1\mu$. Due to this change of μ, the square of the proximity δ^2 (which can be cast as a potential function) might increase to as large as $\mathcal{O}(n)$. For example, if the current iterate is on the central path, then $v = e$ and $\delta^2 = 0$. However, after the update of μ the new proximity becomes $\delta_+^2 = (n/2)(10 - 2 + 0.1) = 4.05n$. In this case, the primal-dual pair (x, s) goes far outside the neighborhood $\mathcal{N}(\mu, \tau)$. Thus we need to initialize the inner procedure to reduce the proximity. It has been proven by several researchers [89,98] that for a suitable step size, $\delta_+^2 - \delta^2 \leq -\beta$, where β is a specific constant. This result implies that the number of inner iterations might be as large as $\mathcal{O}(n)$ to get recentered. Keep in mind that at each outer iteration, we reduce μ by a factor 0.1; hence the number of total outer iterations is bounded above by any constant k satisfying $n\, 0.1^k \mu_0 \leq \varepsilon$. It follows immediately that the number of outer iterations is less than $\mathcal{O}(\log(n\mu_0/\varepsilon))$. Multiplying this number by our estimate of the number of inner iterations between two successive outer iterations, we find that the large-update IPM has an $\mathcal{O}(n \log(n/\varepsilon))$ iteration bound [98,122,127]. This result is worse than that for small-update IPMs. However,

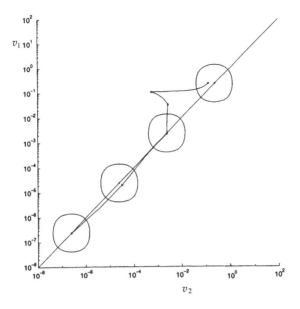

Figure 1.1 Performance of a large-update IPM.

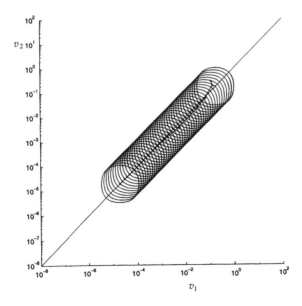

Figure 1.2 Performance of a small-update IPM.

as demonstrated by Figures 1.1 and 1.2, large-update IPMs perform much more efficiently in practice than small-update methods [7]. This is the gap that Renegar [97] pointed out as one irony of interior-point algorithms. Figures 1.1 and 1.2 exhibit the practical performance of an IPM with large-update and small-update for a specific two-dimensional LO problem. These figures are drawn in the v-space.

To close this gap, several remedies have been proposed by various authors aiming at interpreting the practical performance of interior-point algorithms or improving the complexity of large update IPMs. For example, instead of the worst-case analysis, Todd [108] and Ye [125] argued that the average-case iteration bound of large-update IPMs for LO problem instances based on certain probability distribution is $\mathcal{O}(L \log n)$, while small-update IPMs have an average $\mathcal{O}(n^{1/4}L)$ iteration bound. Another attempt to improve the complexity of large-update IPMs was first considered by Monteiro, Adler and Resende [77], who utilized so-called higher-order methods to improve the complexity of large-update IPMs for linear and quadratic optimization problems. Their ideas were further extended and studied for variants of IPMs for solving more complex problems by Hung and Ye [42], Jansen et al. [48], Zhang and Zhang [130]. For IPMs with higher-order correctors, some additional equation systems that provide a higher-order approximation to the system (1.5) must always be solved at each iteration.

Unfortunately neither of these two approaches are very satisfying. Having no other reasonable explanations, most experts in the IPM field then accept unwillingly a vague assertion that small-update IPMs are confined to unacceptably slow progress, while the large-update IPMs are much more adventurous and might have the potential for faster progress in practice.

A natural question arising here is: *Can we lower the complexity bounds of large-update methods by improved analysis without solving any additional systems?*

1.3 PRELIMINARIES AND SCOPE OF THE MONOGRAPH

1.3.1 Preliminary Technical Results

In the last part of this introductory chapter, we first review some of our previous technical results and then describe some intuitive observations that serve to spark the research of the authors. We begin by presenting some mathematical results. This is because these technical results are not only completely independent of the algorithm, but also essential in most of the analysis that we carry out for our new IPMs in the later chapters. Further, these techniques might also take a key position in future developments of the complexity theory of IPMs.

Let us start with a small exercise. Assume that one has a big number, say t_0, which in practice might mean some amount of a material. We consider the decreasing behavior of a sequence starting from this number, or in other words the reduction of the material, since the material will be reduced whenever it is used. The question is: when will the material run out? The answer is instinctive if the material is used at a constant frequency with a fixed amount each time. However, this is not the customary way for most people, since whenever the material is abundant, people are likely to lavish more than necessary until they realize a shortage of the material is imminent. Hence a more convincing way to predict the reduction of the material is to assume that it reduces at a speed that depends on the current situation of the material.

The mathematical model for the above situation can be phrased as follows: starting from a positive number t_0, we decrease it to $t_{k+1} = t_k - f(t_k)$, where f is a specific function. The question is: after how many steps will we obtain $t_k \leq 0$? This simple model resembles the inner iterative process in IPMs. Observe that the total number of iterations of a primal-dual Newton-type algorithm described in the previous section is composed of just two parts: the number of outer and inner iterations. Since we reduce the parameter μ by a constant factor ($\mu_+ = (1 - \theta)\mu$) at each outer iteration, the number of outer iterations follows readily. The main effort in the study of IPMs is to figure out an answer to the following question: how many inner iterations are needed to recenter? Note that during the inner process, we always use corrector steps aimed at minimizing certain proximities. It will be desirable if we can reduce the proximity according to its current value. In other words, if the point is far away from the central path, then we hope to move to the path quickly, depending on the present position of the point. Unfortunately, this is not an easy task. For large-update IPMs, which are clearly the best workhorses in IPM implementations [7], most of the known results, as we describe in Section 1.2.3, show only that the proximity (or the potential function) has at least a constant decrease in each inner iteration, resulting in the worse $\mathcal{O}(nL)$ iteration bounds of the algorithms [98,127]. Obviously if we can prove that the decrease of the proximity after one step has a value associated with the present iterate, for instance $\delta_+^2 \leq \delta^2 - \beta\delta^\gamma$ for some $\beta, \gamma > 0$ (which means when the proximity δ^2 is large, it decreases at a rate much greater than just only a constant decrease), then the complexity of large-update algorithms might be improved. Note that in such a situation, the decreasing behavior of the proximity can be simply described by a specific case of the positive decreasing sequence $t_{k+1} = t_k - f(t_k)$ with $f(t) = \beta t^\gamma$.

Let us consider a concrete positive sequence, say $t_0 = 10000$ and $t_{k+1} = t_k - t_k^{1/2}$. The problem to be addressed is: after how many iterations do we get $t_k \leq 1$? The behavior of this sequence is demonstrated by Figure 1.3 where the straight line shows the decrease of the sequence defined by $t_{k+1} =$

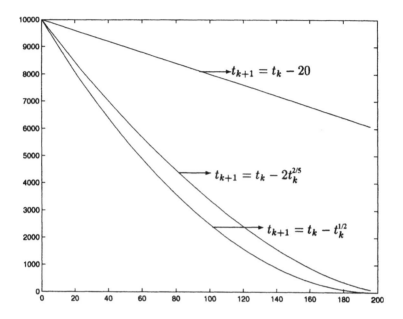

Figure 1.3   The decreasing behavior of three positive sequences.

$t_k - 20$ and another curve illustrates the decreasing behavior of the sequence given by $t_{k+1} = t_k - 2t_k^{2/5}$.

By a simple program, we quickly find an answer to the above question: after 196 steps one gets $t_{197} \leq 1$ for the sequence generated by $t_{k+1} = t_k - t_k^{1/2}$. We also observe that the sequence given by $t_{k+1} - 2t_k^{2/5}$ also reduces to a very small number after 200 steps, while the straight line is still above the horizontal line $t = 6000$. In what follows we give a mathematical explanation pertinent to the decreasing behavior of these positive sequences. First we present a technical result.

Lemma 1.3.1   *Suppose that $\alpha \in [0, 1]$. Then*

$$(1 + t)^\alpha \leq 1 + \alpha t \qquad \forall t \geq -1. \tag{1.10}$$

Proof   Inequality (1.10) is true because for any fixed $\alpha \in [0, 1]$, the function $f(t) = (1 + t)^\alpha - 1 - \alpha t$ is concave for $t \geq -1$ and $f(0) = f'(0) = 0$, thus $f(t) \leq 0$ if $t \geq -1$. $\qquad\square$

Now we are ready to state our main result in this section. This result plays a fundamental role in our complexity analysis of IPMs in the later chapters.

Proposition 1.3.2 *Suppose that* $\{t_k > 0, k = 0, 1, 2, ..., \overline{k}\}$ *is a given sequence satisfying the inequalities*

$$t_{k+1} \le t_k - \beta t_k^{\gamma}, \quad \beta > 0, \quad k = 0, 1, ..., \overline{k} \qquad (1.11)$$

with $\gamma \in [0, 1)$. *Then*

$$\overline{k} \le \left\lceil \frac{t_0^{1-\gamma}}{\beta(1-\gamma)} \right\rceil.$$

Moreover, for any fixed $\rho \ge 0$, $t_k > \rho$ *implies*

$$k \le \left\lceil \frac{t_0^{1-\gamma} - \rho^{1-\gamma}}{\beta(1-\gamma)} \right\rceil.$$

Proof First we note that if $\beta \ge t_0^{1-\gamma}$, then one step is sufficient. Hence we can assume without loss of generality that $0 < \beta < t_0^{1-\gamma}$. Let us further assume that at the present step (kth step) $0 < \beta < t_k^{1-\gamma}$ holds. Then from (1.11) we have

$$t_{k+1}^{1-\gamma} \le (t_k - \beta t_k^{\gamma})^{1-\gamma} = t_k^{1-\gamma}(1 - \beta t_k^{\gamma-1})^{1-\gamma}$$

$$\le t_k^{1-\gamma}(1 - \beta(1-\gamma)t_k^{\gamma-1}) = t_k^{1-\gamma} - \beta(1-\gamma),$$

where the second inequality follows from (1.10). The proposition follows immediately. $\qquad\qquad\square$

We mention that for the preceding example, the theoretical bound on the number of steps given by Proposition 1.3.2 is 199, which is quite close to the true value: 196 steps. This indicates, to some extent, that our theoretical bound is very tight.

We close this section by presenting a technical lemma about the minimal values of certain specific convex functions.

Lemma 1.3.3 *Suppose that* $h(t)$ *is a twice differentiable convex function with*

$$h(0) = 0, \quad h'(0) < 0.$$

Suppose that $h(t)$ *attains its global minimum at its stationary point* $t^* > 0$ *and* $h''(t)$ *is increasing with respect to* t. *Then for any* $t \in [0, t^*]$,

$$h(t) \le \frac{h'(0)t}{2}.$$

Proof Since $h(0) = 0$, for any $t \in [0, t^*]$ we have

$$h(t) = \int_0^t h'(\xi)d\xi = h'(0)t + \int_0^t \int_0^\xi h''(\zeta)d\zeta d\xi \leq h'(0)t + \int_0^t \xi h''(\xi)d\xi$$

$$= h'(0)t + (\xi h'(\xi))|_0^t - \int_0^t h'(\xi)d\xi \leq h'(0)t - h(t),$$

where the first inequality is given by the assumption that $h''(t)$ is nonnegative and increasing with respect to $t > 0$, and the second inequality holds because $h'(t) \leq 0$ for all $t \in [0, t^*]$. The lemma follows directly. $\qquad\square$

Lemma 1.3.3 is elementary in estimating the decrease of our new proximity functions. The decreasing behavior of some concrete forms of the function $h(t)$, such as $h(t) = t - \log(1 + t)$ with $t > -1$, have been well studied in the IPM literature (see [98,127]).

1.3.2 Relation Between Proximities and Search Directions

We proceed with a short review of some ideas and observations that inspired our research. Departing from the probabilistic and higher-order approaches, the authors recently proposed [90,91] a new class of large-update IPMs for solving LO and SDO. To follow the central path more efficiently, we introduced some new search directions that were derived from a new class of proximities for the problem. To describe the algorithms in [90,91] more concretely, let us introduce the following notation:

$$d_x := \frac{v\Delta x}{x}, \quad d_s := \frac{v\Delta s}{s}, \tag{1.12}$$

where v is defined by (1.7). Recall that one can state the centrality condition in (1.5) as $v = v^{-1} = e$. Let us further denote $\overline{A} = \frac{1}{\mu}AV^{-1}X$, $V = \text{diag}(v)$, $X = \text{diag}(x)$. Then we can rewrite system (1.6) in the v-space as

$$\overline{A}d_x = 0,$$

$$\overline{A}^T\Delta y + d_s = 0, \tag{1.13}$$

$$d_x + d_s = v^{-1} - v.$$

Set $d_v = d_x + d_s = v^{-1} - v$. From (1.13) one can easily see that d_x and d_s are the orthogonal decomposition of the vector d_v in the null space and row space of $\overline{A}$, respectively. Another interesting observation is that we can always decompose the above system into two systems. One is the system defining the predictor direction, which is obtained by replacing the last equation in (1.13) by

$$(d_v)_{Pred} = -v,$$

and the other is the system giving the centering direction by

$$(d_v)_{\text{Cent}} = v^{-1}.$$

The predictor direction aims to decrease the duality gap, while the centering direction serves the purpose of centering (it points towards the "analytic center" of the feasible set). It is straightforward to verify that $(d_v)_i \leq 0$ for all the components $v_i \geq 1$ and $(d_v)_i > 0$ for the components $v_i < 1$. This means that if $v_i < 1$ then the classical Newton step increases v_i and decreases it whenever $v_i > 1$ to get closer to the μ-center. It is reasonable to expect that if we can increase the small components and decrease the large components of v more, we might approach our target the μ-center faster, and hence follow the central path more efficiently while staying in a large neighborhood of the path. This straightforward idea is the initial point from which we launched our novel IPM approach for handling classes of problems.

The new search direction, rooted in this simple idea and introduced in [90], is a slight modification of the standard Newton direction. It is defined by a new system as follows:

$$\bar{A}d_x = 0,$$

$$\bar{A}^{\text{T}}\Delta y + d_s = 0 \tag{1.14}$$

$$d_x + d_s = v^{-q} - v, \tag{1.15}$$

where $q \geq 1$ is a parameter. That is, instead of the classical centering direction $(d_v)_{\text{Cent}} = v^{-1}$, we employ a new kind of centering direction such as $(d_v)_{\text{Cent}} = v^{-q}$. At first glance, one can find that compared with the classical Newton search direction, at least locally the new search direction with $q > 1$ will decrease the large elements $v_i > 1$ and increase the small elements $v_i < 1$ more. This further indicates that the new search direction might help us to find a new primal-dual (x, s) close to the central path faster. This is the reason why we were able to improve the complexity of large-update IPMs in [90].

To extract a relatively deeper and mathematically logical interpretation, let us dwell on this system a little more. For instance, consider the special case where $q = 3$. It is clear that the right-hand side of (1.15) represents the negative gradient of the proximity measure δ^2 in the v-space. When solving this system, we get the steepest descent direction for the proximity measure δ^2, along which the proximity can be driven to zero. It is also of interest to note that the right-hand side, $v^{-1} - v$ of the last term in (1.13) is the negative gradient of the scaled classical barrier function Φ, which means the standard primal-dual Newton method is identical to the steepest descent method for minimizing Φ in the scaled v-space. With a deeper look back at the IPM literature, one can find that most potential-reduction methods for solving LO utilize the gradient of the potential barrier function to define a search direction. For instance, the primal-dual potential function in [127] (see Section 4.3) can be rewritten equivalently as

$$\Phi_1(xs, \mu) = (n + q_0) \log \|v\|^2 - \sum_{i=1}^{n} \log v_i^2, \quad q_0 \geq \sqrt{n},$$

while the search direction satisfies

$$d_x + d_s = v^{-1} - \frac{n + q_0}{\|v\|^2} v.$$

It is straightforward to verify that the search direction considered by Ye is proportional to those defined by (1.13) with v replaced by $v \sqrt{n + q_0}/\|v\|$, *which is similar to an update of* μ.

From these observations, we were naturally led to the idea that whenever another proximity measure is used in the algorithm, one should adapt the search direction correspondingly. The algorithms posed in [91] follow such a schedule: they use a special class of potential functions (or proximities) in the path-following method to control the "distance" of the point to the central path, and then the search direction is obtained by solving system (1.6), where the right-hand side of (1.15) is represented by the gradient of the potential function in the scaled v-space. Such an idea can be viewed as an elegant combination of the path-following method and the potential reduction method. The functions introduced in [91] are named *self-regular* functions, a name inherited from the *self-concordant* functions introduced by Nesterov and Nemirovskii [83]. As shown in [91], *self-regular* functions enjoy certain attractive properties not shared by *self-concordant* functions. For instance, as we can also see in Chapter 2, a self-regular function with a *barrier degree* $q > 1$ has a much stronger barrier property than the classical logarithmic barrier function. There exist also several strong relations among a self-regular function and its derivatives. This monograph utilizes the same idea as that in [91], namely we consider the primal-dual problem under the setting of the scaled v-space. Then we define self-regular proximities derived from corresponding functions and use them to keep control on the "distance" from the present iterate to the central path. New search directions are proposed based on those self-regular proximities. By using versatile appealing properties of *self-regular* functions, we develop some new IPMs with large update for solving classes of problems, and show that the complexity of these new algorithms can get arbitrarily close to the best known iteration bounds of IPMs.

1.3.3 Contents and Notational Abbreviations

This work aims at setting up a unified framework for the complexity theory of a class of primal-dual path-following methods for various problems including linear optimization (LO), second-order cone optimization (SOCO), semidefinite optimization (SDO), and nonlinear complementarity problems (NCPs). The work consists of eight chapters. In Chapter 2, we first introduce the definition of self-regular functions in $\Re_{++}$. Then we investigate several important properties of self-regular functions such as the growth behavior, the barrier behavior and the relations among the function and its first and second derivatives. Several generating rules to construct a self-regular function are also discussed. Common features and differences between self-regular functions and the well-known self-concordant functions are discussed. The content of this chapter consists of several conclusions from [89,91] but we refine and reorganize them slightly differently here. We also postpone the introduction of self-regular functions in semidefinite and second-order cones to the corresponding chapters where the problem under consideration is also described in more detail. It may be possible to build up a unified framework for self-regular functions depending on some abstract mathematical theory, but we think that first a simple and concrete introduction and then its more complex evolution will be helpful for most people to understand the concept of *self-regularity* more easily. We also expect that such an arrangement will benefit readers who are interested in just one part of this work, the case of LO.

In Chapter 3, we first present the notion of *self-regular proximity* for LO, which is derived from the self-regular function. Self-regular proximity for LO is defined in the so-called v-space, which we described earlier in Section 1.2.3. The *growth behavior* of the proximity as well as its *barrier property* are studied.[6] Then we introduce some new search directions for path-following algorithms for LO. Using the properties of the proximity, we prove that the large-update path-following method for LO has a polynomial $\mathcal{O}(n^{(q+1)/(2q)} \log(n/\varepsilon))$ iteration bound, where $q \geq 1$ is the so-called *barrier degree* of the proximity, and the constant hidden in the expression can be defined as a function of the barrier degree q, *the growth degree p* of the proximity and two other constants ν_1 and ν_2 related only to the proximity. When q increases, our result approaches the best known complexity $\mathcal{O}(\sqrt{n} \log(n/\varepsilon))$ iteration bound for interior-point methods. Our unified analysis also provides the $\mathcal{O}(\sqrt{n} \log(n/\varepsilon))$ complexity for small-update IPMs. At each iteration, we need to solve only one linear system. The possibility of relaxing some conditions is also explored.

Chapter 4 depicts a direct extension of the algorithms posed in Chapter 3. The class of complementarity problems (CPs) is chosen as our underlying problem. Many problems in mathematical programming can be reformulated as CPs. One can list problems such as LO, QP and general convex program-

---

[6] Throughout this monograph, the growth behavior of a function $f(t)$ is the change of the function as its argument t goes to infinity. If the value of $f(t)$ tends to infinity as t reduces to zero, then we say $f(t)$ has a barrier property. The growth degree and barrier degree of a function are defined in Chapter 2.

ming problems that satisfy certain constraint qualifications. IPMs for solving CPs try to find a primal-dual pair (x, s) in the nonnegative orthant such that both a nonlinear equation[7] and the complementarity condition $xs = 0$ are satisfied. Because of the appearance of nonlinearity, only a specific class of CPs, which always require certain sufficient smoothness conditions (e.g., the scaled Lipschitz condition, or the self-concordancy condition), can be solved efficiently by IPMs. In this chapter, some new smoothness conditions are introduced. We show that, if the considered problem satisfies our new condition, then our new primal-dual path-following method has the same polynomial iteration bounds as its LO analogue.

Having built up the new framework of primal-dual IPMs for LO and NCPs, in Chapter 5 we discuss the generalization of such algorithms to SDO, where the complementarity condition is expressed by $XS = 0$ with both X and S in the positive semidefinite cone $S_+^{n \times n}$. An important feature for SDO is that, to make the Newton system solvable in the space of $n \times n$ symmetric matrices $S^{n \times n}$, one has to apply some scaling scheme so that the rescaled Newton system has necessarily a unique symmetric solution. For this we first discuss the proximities based on different scaling techniques. After exploiting some fascinating features of self-regular functions on the positive definite cone $S_{++}^{n \times n}$, we show that, among others, the NT-scaling is the unique optimal choice if the corresponding proximity is self-regular. New search directions for SDO are then proposed rooted in the NT-scaling and self-regular proximities in $S_{++}^{n \times n}$. Certain functions on the cone of general symmetric matrices are discussed and their first and second derivatives are estimated. To the authors' best knowledge, this is the first time that estimates have been obtained for the derivatives of a general function involving matrix functions. The results are meaningful also in a pure mathematical sense. They might be very helpful in the future study of general nonlinear optimization with matrix variables. Finally the complexity of the algorithm is built up on the aforementioned results.

SOCO is a specific class of problems lying between LO and SDO. Theoretically the solution of a SOCO problem can be obtained by finding the optimal solution of an SDO problem relevant to the original SOCO problem. However, for this one usually has to face both a large increase in the problem's size and the loss of certain advantages of the original problem. Thus, with regard to IPMs, the algorithm working on the second-order cone is a favorable choice for solving SOCO. To build up the complexity of the algorithm for SOCO, in Chapter 6 we first explore versatile properties of the functions associated with the second-order cone and estimate their first and second directional derivatives. As a byproduct, these estimates also verify some of our conclusions

---

[7] For LO and LCPs, we only need to identify a primal-dual pair (x, s) satisfying a linear system of equations and the complementarity conditions.

about matrix functions in Chapter 5. Particular attention is paid to self-regular functions and self-regular proximities on second-order cones. Then, based on these self-regular proximities, new search directions are suggested. It is shown that the algorithm for SOCO has the same complexity as its analogue for LO. Let us emphasize again that only one linear system needs to be solved at each iteration.

Chapter 7 describes a variety of homogeneous models for classes of problems considered in this monograph. These models have been well studied in the IPM literature and are widely used to embed the original problem into a slightly augmented one for which a strictly feasible starting point can be easily obtained so that feasible IPMs can be applied directly. Further, necessary information about the solution of the primitive issue can be obtained from a solution of the new problem.

In Chapter 8, we close this work by listing some concluding remarks and suggestions on the implementation issues pertinent to our algorithms. Several topics for future research are also addressed.

We conclude this introductory chapter with a few words about the notation. As mentioned earlier, this context covers numerous topics ranging from the simplest LO over polyhedrons to LO over more complex conic constraints; from a problem of minimizing an affine objective to that with a highly nonlinear objective. This unfortunately results in a very heavy notational effort. Although we have listed most of the notation before we started this chapter, readers should be aware that sometimes for consistency of mathematical expression, we use the same notation for different arguments. For instance, $\psi(t)$ denotes a univariate function while $\psi(x)$ is a mapping from $\Re^n$ (or the second-order cone K) into itself. Moreover, for any symmetric matrix X, we denote by $\psi(X)$ a mapping from $S^{n \times n}$ into itself. Also the notation $\mathcal{N}(\tau, \mu)$ has distinct meanings in separate chapters. We refer the readers to the corresponding chapters for details about such definitions.

Chapter 2

Self-Regular Functions and Their Properties

This chapter presents a comprehensive study about a new class of functions: the univariate self-regular functions in $\Re_{++}$. After a basic introduction, several generating rules for constructing self-regular functions are discussed. Some appealing properties of self-regular functions are investigated, including the growth behavior, the barrier behavior and the relations among a self-regular function and its first and second derivatives. Most of these results are used in the next chapter, where we show that self-regular functions can be used to improve the complexity bound for large-update IPMs. The common features and differences between self-regular and self-concordant functions are also addressed. This chapter forms the bedrock of the notion of self-regularity, which we later develop further in different spaces.[1]

---

[1] Readers who are not interested in the complexity theory of IPMs may skip this technical section at the first reading and move to the algorithmic part in later chapters.

2.1 AN INTRODUCTION TO UNIVARIATE SELF-REGULAR FUNCTIONS

As already mentioned earlier in Section 1.3.2, proximity measures or potential functions play an important role in both the theoretical study and the practical implementation of IPMs. From the outset, the pioneering work of Karmarkar [52] employed a potential function to keep the iterative sequence in the interior of the feasible set. The sharp observation by Gill et al. [28] made it clear that Karmarkar's potential function is essentially related to the classical *logarithmic barrier function* [23]. This led to a rebirth of some classical algorithms for nonlinear programming.

It was the discovery of the central path (and the related concept of the analytic center of a convex set) by Sonnevend [102] and Megiddo [70] that inaugurated a new era for IPMs. Since then, it is reasonable to claim that most IPMs have originated from the same idea, to follow various central paths appropriately to the optimal set. This can be the central path in the primal space, or in the dual space, or in the primal-dual space with certain strategies. It is not surprising to see that most IPMs are closely related to some older methods that gave rise to the central path: various logarithmic barrier functions in suitable spaces.

The development of IPMs based on the logarithmic barrier approach reached a peak when Nesterov and Nemirovskii set up the prototype of certain variants of IPMs for solving a wide class of problems in [83], where various barriers for different settings are designed. The framework in [83] is built upon the theory of *self-concordance*.

In this work, we try to build up a new paradigm for primal-dual IPMs on the bedrock of *self-regularity* and eliminate the irony of Renegar. We start by considering the standard LO problem and an algorithm based on the primal-dual setting. We are particularly interested in the construction of proximities and search directions in primal-dual path-following methods that can be applied to trace the central path more efficiently, at least from a theoretical point of view.

Recall that, as we described in Sections 1.2.3 and 1.3.2, by using the so-called v-space, we can rewrite the centrality condition as $v = e$ or, equivalently coordinatewise, $v_i = 1$ for all $i = 1, 2, \ldots n$. Hence, taking it gently, we can consider consequently a univariate function in $\Re_{++}$ that attains its global minimum at 1 and can be used to measure the distance from any point in $\Re_{++}$ to 1. This is the primary goal of the new function. Moreover, it is desirable for the function to enjoy a certain barrier property that prevents the argument from moving to the boundary of $\Re_{++}$. This partly explains the conditions that are introduced below to define the notion of a univariate *self-regular* function. As we can see in later chapters, the concept of self-regular functions can be transparently extended to the cases of various cones such as the positive orthant, the second-order cone, and the cone of positive definite matrices.

Definition 2.1.1 *The function* $\psi(t) \in C^2 : (0, \infty) \to \Re$ *is self-regular if it satisfies the following conditions:*

SR1 *$\psi(t)$ is strictly convex with respect to $t > 0$ and vanishes at its global minimal point $t = 1$, that is, $\psi(1) = \psi'(1) = 0$. Further, there exist positive constants $\nu_2 \geq \nu_1 > 0$ and $p \geq 1, q \geq 1$ such that*

$$\nu_1(t^{p-1} + t^{-1-q}) \leq \psi''(t) \leq \nu_2(t^{p-1} + t^{-1-q}) \quad \forall t \in (0, \infty); \quad (2.1)$$

SR2 *For any $t_1, t_2 > 0$,*

$$\psi(t_1^r t_2^{1-r}) \leq r\psi(t_1) + (1-r)\psi(t_2), \quad r \in [0, 1]. \quad (2.2)$$

We call parameter q the *barrier degree* and p the *growth degree* of $\psi(t)$, if it is self-regular. In the rest of this section we consider some immediate consequences of the above definitions. In subsequent sections we consider some other properties of self-regular functions that are relevant in the analysis of interior-point methods for LO.

Obviously, if $\psi(t)$ is self-regular, then so is $\psi(t)/\nu_1$. Therefore, without loss of generality, for the moment we may assume $\nu_1 = 1$.

Note that an immediate consequence of $\psi(1) = \psi'(1) = 0$ is that

$$\psi(t) = \int_1^t \int_1^\xi \psi''(\zeta) d\zeta d\xi. \quad (2.3)$$

Now consider the special case where $\nu_1 = \nu_2 = 1$. Then (2.1) implies

$$\psi''(t) = t^{p-1} + t^{-1-q},$$

and, as a consequence, $\psi(t)$ is uniquely determined by p and q. The unique self-regular function obtained in this way plays an important role in the sequel. It will be denoted as $\Upsilon_{p,q}(t)$. By integrating twice, one may easily verify that, with $p \geq 1$,

$$\Upsilon_{p,1}(t) = \frac{t^{p+1} - 1}{p(p+1)} - \frac{\log t}{q} + \frac{p-1}{p}(t-1), \quad (2.4)$$

$$\Upsilon_{p,q}(t) = \frac{t^{p+1} - 1}{p(p+1)} + \frac{t^{1-q} - 1}{q(q-1)} + \frac{p-q}{pq}(t-1), q > 1. \quad (2.5)$$

Since

$$\lim_{q \to 1} \frac{t^{1-q} - 1}{q-1} = -\log t,$$

we have

$$\lim_{q \to 1} \Upsilon_{p,q}(t) = \Upsilon_{p,1}(t).$$

In view of this relation, all the results below in terms of $\Upsilon_{p,q}(t)$ (with $q > 1$) can be extended to the case $q = 1$, by continuity. It may be worthwhile to point out that

$$\Upsilon_{1,1}(t) := \frac{t^2 - 1}{2} - \log t$$

is the well-known (univariate) logarithmic barrier function [98].

Figure 2.1 demonstrates the growth and barrier behaviors of two functions $\Upsilon_{1,1}(t)$ and $\Upsilon_{1,3}(t)$. As we can see from this figure, the growth behaviors of these two functions as $t \to \infty$ are quite similar. However, when $t \to 0$, the function $\Upsilon_{1,3}(t)$ has a much stronger barrier property than $\Upsilon_{1,1}(t)$.

As a consequence of the above definitions, condition (2.1) can be equivalently stated as

$$\nu_1 \Upsilon_{p,q}''(t) \le \psi''(t) \le \nu_2 \Upsilon_{p,q}''(t). \tag{2.6}$$

Using this relation, one can further prove that for any function $\psi(t)$ satisfying SR1, there holds

$$\nu_1 \Upsilon_{p,q}(t) \le \psi(t) \le \nu_2 \Upsilon_{p,q}(t), \tag{2.7}$$

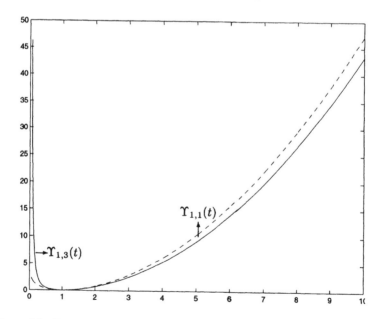

Figure 2.1   Demonstration of self-regular functions.

and, using that the derivatives of $\psi(t)$ and $Y_{p,q}(t)$ are positive if $t > 1$ and negative if $t < 1$, one can easily prove that

$$\nu_1|Y_{p,q}'(t)| \le |\psi'(t)| \le \nu_2|Y_{p,q}'(t)|. \tag{2.8}$$

At this stage it is not yet clear that the functions $Y_{p,q}$ are self-regular. By definition, condition SR1 is satisfied. Below we show that $Y_{p,q}$ also satisfies SR2.

To facilitate our analysis, let us denote by Ω_1 and Ω_2 the sets of functions whose elements satisfy conditions SR1 and SR2, respectively.[2]

We start with an important lemma[3] that gives a different characterization of condition SR2.

Lemma 2.1.2 *A function* $\psi(t) \in C^2 : (0, +\infty) \to \Re$ *belongs to* Ω_2 *if and only if the function* $\psi(\exp(\zeta)) : \Re \to \Re$ *is convex in* ζ, *or equivalently* $\psi'(t) + t\psi''(t) \ge 0$ *for* $t > 0$.

Proof The proof is straightforward. First from the definition of convexity, we know that $\psi(\exp(\zeta))$ is convex if and only if for any $\zeta_1, \zeta_2 \in \Re$, the following inequality holds:

$$\psi(\exp(r\zeta_1 + (1 - r)\zeta_2)) \le r\psi(\exp(\zeta_1)) + (1 - r)\psi(\exp(\zeta_2)), \quad r \in [0, 1]. \tag{2.9}$$

Letting $t_1 = \exp(\zeta_1)$, $t_2 = \exp(\zeta_2)$, obviously one has $t_1, t_2 \in (0, +\infty)$. Further, relation (2.9) can be equivalently rewritten as

$$\psi(t_1^r t_2^{1-r}) \le r\psi(t_1) + (1 - r)\psi(t_2), \quad r \in [0, 1].$$

Since $\psi(\exp(\zeta))$ is convex, that is, $\exp(2\zeta)\psi''(\exp(\zeta)) + \exp(\zeta)\psi'(\exp(\zeta)) \ge 0$, substituting $t = \exp(\zeta)$ gives $t\psi'(t) + t^2\psi''(t) \ge 0$, which is equivalent to $\psi'(t) + t\psi''(t) \ge 0$ for $t > 0$. $\qquad\square$

Note that condition SR2 can be restated equivalently (it is a standard exercise in convexity theory) as

$$\psi(t_1 t_2) \le \frac{1}{2}\left(\psi(t_1^2) + \psi(t_2^2)\right) \quad \forall t_1, t_2 > 0. \tag{2.10}$$

Using Lemma 2.1.2, one gets the technical result.

---

[2] Condition SR1 is a very simple condition that can be verified without much difficulty. Hence in this section we mainly focus on Condition SR2.

[3] Another reason why we give such a technical lemma is that the second condition in the lemma has been widely used in the analysis of singular value inequalities in matrix theory [40]. However, our original Condition SR2 is clearer and more suitable for our specific purpose.

Lemma 2.1.3 *Suppose that $\psi(t) \in \Omega_2$. Then for any $\alpha \in \Re$, $\psi(t^\alpha) \in \Omega_2$.*

Proof If $\alpha = 0$, then the statement of the lemma holds trivially. Therefore, it suffices to discuss the case $\alpha \neq 0$. Let $\psi_1(t) = \psi(t^\alpha)$. By simple calculus we obtain

$$\psi_1'(t) = \alpha t^{\alpha-1} \psi'(t^\alpha), \quad \psi_1''(t) = \alpha(\alpha-1)t^{\alpha-2}\psi'(t^\alpha) + \alpha^2 t^{2\alpha-2}\psi''(t^\alpha).$$

It follows from the assumption of the lemma that

$$\psi_1'(t) + t\psi_1''(t) = \alpha^2 t^{\alpha-1}(\psi'(t^\alpha) + t^\alpha \psi''(t^\alpha)) \geq 0 \quad \forall t > 0.$$

Now by using Lemma 2.1.2, we get the desired conclusion. $\square$

It is straightforward to verify that the functions $\log t$ and $-\log t$ are in Ω_2, and also $t^\rho \in \Omega_2$ for any $\rho \in \Re$. Therefore we readily obtain the following proposition.

Proposition 2.1.4 *Let N be any nonnegative integer. Then for any $\beta_0 \in \Re$, the function*

$$\psi(t) = \beta_0 \log t + \sum_{i=1}^{N} \beta_i(t^{\rho_i} - 1), \quad \beta_i \geq 0, \ \rho_i \in \Re, \ i = 1, 2, ..., N \quad (2.11)$$

belongs to Ω_2.

Hence any function $\psi(t)$ defined by (2.11) that satisfies condition SR1 is *self-regular*.

Now let us return to consider the special case $\Upsilon_{p,q}(t)$.

Lemma 2.1.5 *The function $\Upsilon_{p,q}(t)$ is self-regular when $p, q \geq 1$.*

Proof It suffices to show that $\Upsilon_{p,q}(t) \in \Omega_2$. Through simple calculus, one has

$$\Upsilon_{p,q}'(t) + t\Upsilon_{p,q}''(t) = \frac{p+1}{p}t^p + \frac{q-1}{q}t^{-q} + \frac{1}{q} - \frac{1}{p}.$$

Hence, it follows immediately that for any $p \geq q \geq 1$,

$$\Upsilon_{p,q}'(t) + t\Upsilon_{p,q}''(t) > \frac{1}{q} - \frac{1}{p} \geq 0.$$

Thus, by Lemma 2.1.2, it remains to prove that

$$\Upsilon_{p,q}'(t) + t\Upsilon_{p,q}''(t) \geq 0 \quad \forall q > p \geq 1, \ t > 0.$$

Note that when $q > p \geq 1$, one has $1 - 1/q \geq 1/p - 1/q$, which further implies that for any $t > 0$,

$$Y'_{p,q}(t) + tY''_{p,q}(t) = \frac{p+1}{p}t^p + \frac{q-1}{q}t^{-q} + \frac{1}{q} - \frac{1}{p}$$

$$\geq \frac{p+1}{p}t^p + \frac{q-p}{pq}t^{-q} + \frac{1}{q} - \frac{1}{p}$$

$$> \left(\frac{1}{p} - \frac{1}{q}\right)(t^p + t^{-q} - 1) > 0. \qquad \square$$

To show that this is not the only family of self-regular functions, we proceed with another example.

Example 2.1.6 Consider the function

$$\psi(t) = \frac{t^{p+1} - 1}{p+1} + \frac{t^{1-q} - 1}{q-1}, \quad p \geq 1, \ q > 1,$$

for which

$$\psi'(t) = t^p - t^{-q}$$

$$\psi''(t) = pt^{p-1} + qt^{-q-1}.$$

It is obvious that condition SR1 is satisfied (with $\nu_1 = \min(p, q)$, $\nu_2 = \max(p, q)$). To check SR2, we use Lemma 2.1.2:

$$\psi'(t) + t\psi''(t) = (p+1)t^p + (q-1)t^{-q} \geq 0.$$

This shows that $\psi(t)$ is self-regular. Note that if $p = q$ then $\psi(t) = pY_{p,p}(t)$, but otherwise the functions $\psi(t)$ and $Y_{p,q}(t)$ are linearly independent.

A very simple but quite useful observation is that any nontrivial nonnegative combination of two self-regular functions ψ_1 and ψ_2 is still self-regular.

Proposition 2.1.7 *If the functions* $\psi_1(t), \psi_2(t)$ *are self-regular, then any combination* $\beta_1\psi_1 + \beta_2\psi_2$ *with* $\beta_1, \beta_2 \geq 0$, $\beta_1 + \beta_2 > 0$ *is self-regular.*

Proof We need only consider the case where both β_1 and β_2 are positive. Since the functions $\psi_1(t), \psi_2(t) \in \Omega_1 \cap \Omega_2$, it follows immediately from Lemma 2.1.2 that the function $\psi(t) := \beta_1\psi_1(t) + \beta_2\psi_2(t) \in \Omega_2$. Obviously $\psi(t)$ is strictly convex and has a global minimizer at $t = 1$. Hence it remains to show that the imposed condition (2.1) with respect to the second derivative $\psi''(t)$ holds for some constants $\nu_1, \nu_2 > 0$ and $p, q \geq 1$. Since both $\psi_1(t)$ and

$\psi_2(t)$ are self-regular, there exist constants $\nu_1^1, \nu_2^1, \nu_1^2, \nu_2^2 > 0$ and $p_1, q_1, p_2, q_2 \geq 1$ such that

$$\nu_1^1(t^{p_1-1} + t^{-1-q_1}) \leq \psi_1''(t) \leq \nu_2^1(t^{p_1-1} + t^{-1-q_1}), \tag{2.12}$$

$$\nu_1^2(t^{p_2-1} + t^{-1-q_2}) \leq \psi_2''(t) \leq \nu_2^2(t^{p_2-1} + t^{-1-q_2}). \tag{2.13}$$

Let $p = \max(p_1, p_2)$, $q = \max(q_1, q_2)$. Obviously $p, q \geq 1$ holds. From the definition of $\psi(t)$ we obtain

$$\psi''(t) \geq \beta_1 \nu_1^1(t^{p_1-1} + t^{-1-q_1}) + \beta_2 \nu_1^2(t^{p_2-1} + t^{-1-q_2})$$

$$= (\beta_1 \nu_1^1 t^{p_1-1} + \beta_2 \nu_1^2 t^{p_2-1}) + (\beta_1 \nu_1^1 t^{q_1-1} + \beta_2 \nu_1^2 t^{q_2-1})$$

$$> \min(\beta_1, \beta_2) \min(\nu_1^1, \nu_1^2)(t^{p-1} + t^{-1-q}).$$

Note that from the choice of p and q, we have

$$0 < \frac{t^{p_1-1} + t^{-1-q_1}}{t^{p-1} + t^{-1-q}} = \frac{t^{p_1-1}}{t^{p-1} + t^{-1-q}} + \frac{t^{-1-q_1}}{t^{p-1} + t^{-1-q}} < 2,$$

$$0 < \frac{t^{p_2-1} + t^{-1-q_2}}{t^{p-1} + t^{-1-q}} = \frac{t^{p_2-1}}{t^{p-1} + t^{-1-q}} + \frac{t^{-1-q_2}}{t^{p-1} + t^{-1-q}} < 2.$$

This relation, combining with (2.12) and (2.13), yields

$$\psi''(t) \leq 2(\beta_1 \nu_2^1 + \beta_2 \nu_2^2)(t^{p-1} + t^{-1-q}).$$

Let us choose

$$\nu_1 = \min(\beta_1, \beta_2) \min(\nu_1^1, \nu_1^2), \quad \nu_2 = 2(\beta_1 \nu_2^1 + \beta_2 \nu_2^2).$$

From our discussions above, we can claim that

$$\nu_1(t^{p-1} + t^{-1-q}) \leq \psi''(t) \leq \nu_2(t^{p-1} + t^{-1-q}),$$

which completes the proof of the proposition. $\qquad\qquad\qquad\qquad\qquad\square$

Proposition 2.1.7 means that the set of self-regular functions is a pointed convex cone.

We proceed to give another generating rule for self-regular functions. As asserted earlier in Lemma 2.1.3, if $\psi(t) \in \Omega_2$, so is the function $\psi(1/t)$. In what follows we consider a converse question. Suppose a function satisfies the relation $\psi(t) = \psi(1/t)$. The issue we want to address is when such a function is self-regular. Our next result gives a positive answer to this question when $\psi(t) \in \Omega_1$, and thus provides another way to generate self-regular functions. This result is used later in our discussions about the relations between self-regular functions and self-concordant functions.

Lemma 2.1.8 *If $\psi(t) = \psi(t^{-1})$ and $\psi(t) \in \Omega_1$, then $\psi(t)$ is self-regular.*

Proof It suffices to prove that for any $t_1, t_2 > 0$,

$$\psi(t_1 t_2) \le \frac{1}{2}\left(\psi(t_1^2) + \psi(t_2^2)\right).$$

Since $\psi(t) \in \Omega_1$, $\psi(t)$ is increasing for $t \ge 1$. Hence if $t_1 t_2 \ge 1$, then by condition SR1,

$$\psi(t_1 t_2) \le \psi\left(\frac{t_1^2 + t_2^2}{2}\right) \le \frac{1}{2}\left(\psi(t_1^2) + \psi(t_2^2)\right).$$

Thus it remains to consider the case $t_1 t_2 < 1$, or in other words $1/t_1 t_2 > 1$. Now using the assumption $\psi(t) = \psi(t^{-1})$, we obtain

$$\psi(t_1 t_2) = \psi\left(\frac{1}{t_1 t_2}\right) \le \frac{1}{2}\left(\psi\left(\frac{1}{t_1^2}\right) + \psi\left(\frac{1}{t_2^2}\right)\right) = \frac{1}{2}\left(\psi(t_1^2) + \psi(t_2^2)\right)$$

as required. $\square$

By way of illustration, consider $\psi(t) = 1/2(t - t^{-1})^2$, a function well known in the IPM literature. Clearly, this function satisfies the hypothesis of Lemma 2.1.8. One may easily check that it satisfies SR1, with $p = \nu_1 = 1$ and $q = \nu_2 = 3$. By Lemma 2.1.8, the self-regularity follows. Note that this function can also be put in the form given by Example 2.1.6.

We close this section by giving some examples that show the independence of conditions SR1 and SR2. Let $\psi_1(t) = t^2 - t - \log t$. It satisfies condition SR1, but, by simple calculus,

$$\psi_1'(t) + t\psi_1''(t) = 4t - 1.$$

From Lemma 2.1.2 we know that $\psi_1(t)$ does not satisfy SR2 because $\psi_1'(t) + t\psi_1''(t) < 0$ whenever $t < 1/4$. On the other hand, the simple function t^2 satisfies SR2 but not SR1. Thus the conditions SR1 and SR2 are independent.

2.2 BASIC PROPERTIES OF UNIVARIATE SELF-REGULAR FUNCTIONS

This section is devoted to investigating various fascinating properties of a self-regular function $\psi(t)$. Most of the results in this section are indeed established under only condition SR1. As we see later in section 3.5, in the case of LO, by using several technical conclusions in this section one can build up the polynomial complexity of the algorithm without resorting to condition SR2.

We start by presenting several intrinsic features of functions $\psi(t) \in \Omega_1$ with $q > 1$. These features exhibit some intriguing relationships among functions $\psi(t) \in \Omega_1$ and their derivatives. Since condition SR1 is a necessary condition for the *self-regularity*, these attractive features are naturally shared by general univariate self-regular functions.

We proceed with a discussion about the relations between a function $\psi(t)$ and its first and second derivatives when $\psi(t) \in \Omega_1$.

Lemma 2.2.1   *Suppose that* $\psi(t) \in \Omega_1$. *Then*

$$\left| \frac{1}{t} \psi'(t) \right| \leq \frac{\nu_2}{\nu_1} \psi''(t), \quad t > 0.$$

Proof   Recall from (2.6) that $\nu_1 \Upsilon''_{p,q}(t) \leq \psi''(t)$, and from (2.8) that $|\psi'(t)| \leq \nu_2 |\Upsilon'_{p,q}(t)|$. Hence it suffices to show that

$$|\Upsilon'_{p,q}(t)| \leq t \Upsilon''_{p,q}(t), \quad t > 0. \tag{2.14}$$

By Lemma 2.1.2, $\Upsilon_{p,q}(t) \in \Omega_2$. This means that

$$\Upsilon'_{p,q}(t) + t \Upsilon''_{p,q}(t) = t^p \left(1 + \frac{1}{p} \right) - \frac{1}{p} + t^{-q} \left(1 - \frac{1}{q} \right) + \frac{1}{q} \geq 0. \tag{2.15}$$

Hence it remains to prove that

$$-\Upsilon'_{p,q}(t) + t \Upsilon''_{p,q}(t) \geq 0, \quad t > 0.$$

By simple calculus we can restate this inequality as

$$-\frac{t^p - 1}{p} + \frac{t^{-q} - 1}{q} + t^p + t^{-q} \geq 0,$$

or equivalently,

$$t^p \left(1 - \frac{1}{p} \right) + \frac{1}{p} + t^{-q} \left(1 + \frac{1}{q} \right) - \frac{1}{q} \geq 0, \quad t > 0, \ p \geq 1, \ q \geq 1. \tag{2.16}$$

Replacing t by $1/t$, we see this is equivalent to

$$t^{-p} \left(1 - \frac{1}{p} \right) + \frac{1}{p} + t^q \left(1 + \frac{1}{q} \right) - \frac{1}{q} \geq 0, \quad t > 0, \ p \geq 1, \ q \geq 1.$$

Interchanging p and q we observe that this is exactly the same inequality as (2.15). Hence the proof is complete.   □

Recall that by Lemma 2.1.2, a function $\psi(t) \in \Omega_2$ if and only if $\psi'(t) + t\psi''(t) \geq 0$ for any $t > 0$. For the function $\Upsilon_{p,q}(t)$, clearly $\nu_1 = \nu_2 = 1$ holds. It follows from Lemma 2.2.1 that $\Upsilon'_{p,q}(t) + t \Upsilon''_{p,q}(t) \geq 0$, which is exactly the same as what we have proved in the proof of Lemma 2.1.5.

Proposition 2.2.2 *Suppose that the function $\psi(t)$ satisfies condition SR1 with $q > 1$. Then there is a constant C_ν depending on the constants ν_1, ν_2, p and q such that*

$$\psi(t)\psi''(t) \le C_\nu \psi'(t)^2 \quad \forall t > 0.$$

Proof We first prove the proposition for a special case[4] when $\psi(t) = Y(t) := Y_{p,q}(t)$. In other words, we want to show that there is a constant C_ν such that

$$Y(t)Y''(t) \le C_\nu Y'(t)^2 \quad \forall t > 0.$$

The above inequality holds trivially if $t = 1$. Hence it suffices to show that there exists a constant C_ν satisfying

$$\mathfrak{F}(t) = \frac{Y(t)Y''(t)}{Y'(t)^2} \le C_\nu \quad \forall 1 \ne t > 0.$$

From the definition of $Y(t)$, it easily follows that

$$\lim_{t \to \infty} \mathfrak{F}(t) = \frac{p}{p+1}, \quad \lim_{t>0, t \to 0} \mathfrak{F}(t) = \frac{q}{q-1}.$$

Moreover, by using l'Hospital's rule, one readily obtains

$$\lim_{t \to 1} \mathfrak{F}(t) = \lim_{t \to 1} \frac{Y(t)Y''(t)(t-1)^2}{(t-1)^2 \; Y'(t)^2} = 1.$$

From the above three limits and the continuity of $\mathfrak{F}(t)$ on the intervals $(0, 1)$ and $(1, \infty)$, we can conclude that there exists a constant C_ν satisfying $\mathfrak{F}(t) \le C_\nu$, which matches the statement of the proposition when $\psi(t) = Y(t)$.

Now we consider the case where $\psi(t)$ is a general function satisfying SR1. By (2.1) one has

$$\nu_1 Y''(t) \le \psi''(t) \le \nu_2 Y''(t).$$

Since $\psi'(t) \ge 0$ if $t \ge 1$ and $\psi'(t) < 0$ whenever $t < 1$, from the above relation we obtain

$$0 \le \nu_1 Y'(t) \le \psi'(t) \le \nu_2 Y'(t) \quad \forall t \ge 1,$$

$$\nu_2 Y'(t) \le \psi'(t) \le \nu_1 Y'(t) < 0 \quad \forall t \in (0, 1).$$

It follows that

$$\psi'(t)^2 \ge \nu_1^2 Y'(t)^2 \ge \frac{\nu_1^2}{C_\nu} Y(t)Y''(t) \ge \frac{\nu_1^2}{C_\nu \nu_2^2} \psi(t)\psi''(t),$$

---

[4] For simplicity, we temporarily drop the indices in the function $Y_{p,q}(t)$ and denote it as $Y(t)$.

where the second inequality follows from our discussion for the special case $\psi(t) = \Upsilon(t)$, and the last inequality is a consequence of (2.1) and (2.7). Replacing C_ν by

$$C_\nu := \frac{\nu_2^2}{\nu_1^2} C_\nu,$$

one gets the desired inequality in the proposition. □

Note that the above proposition is not true if $q = 1$. This is demonstrated by Figure 2.2. We also notice that

$$\frac{\psi(t)\psi''(t)}{\psi'(t)^2} = \left(-\frac{\psi(t)}{\psi'(t)}\right)' + 1.$$

Hence Proposition 2.2.2 can be used to bound the derivative of the function $\psi(t)/\psi'(t)$.

The above relations among the function $\psi(t)$ and its first and second derivatives given by Lemma 2.2.1 and Proposition 2.2.2 provide a partial explanation for the name *self-regularity*. As mentioned earlier, we will prove that self-regular functions can be used to improve the complexity of large-update IPMs. Unfortunately we have not been able to use the above properties for that goal. The results that follow in this section are therefore more important: they play a crucial role in the analysis of large-update IPMs.

Now we give some bounds on a function $\psi(t)$ in Ω_1 in terms of t and $\psi'(t)$.

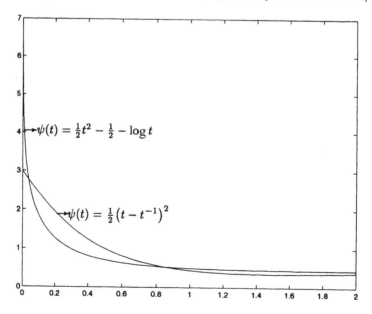

Figure 2.2 Demonstration of the functions defined by $\psi(t)\psi''(t)/\psi'(t)^2$.

Lemma 2.2.3 *Suppose that $\psi(t) \in \Omega_1$. Then*

$$\frac{1}{2}(t-1)^2 \le \frac{\psi(t)}{\nu_1}, \tag{2.17}$$

$$\frac{t^{p+1}-1}{p(p+1)} - \frac{t-1}{p} \le \frac{\psi(t)}{\nu_1}, \tag{2.18}$$

$$\frac{t^{1-q}-1}{q(q-1)} + \frac{t-1}{q} \le \frac{\psi(t)}{\nu_1}, \tag{2.19}$$

$$\psi(t) \le \frac{1}{2\nu_1}\psi'(t)^2. \tag{2.20}$$

Proof The first three inequalities (2.17)–(2.19) are immediate consequences of the following relations:

$$\psi''(t) \ge \nu_1(t^{p-1} + t^{-q-1}) \ge \nu_1 \max\left\{t^{p-1}, t^{-q-1}\right\}$$

$$= \nu_1 \max\left\{1, t^{p-1}, t^{-q-1}\right\} \quad \forall t > 0, \; p, q \ge 1.$$

Using condition SR1, one can easily see that

$$\frac{\psi''(t)}{\nu_1} > 1.$$

It follows that

$$\psi(t) = \int_1^t \int_1^\xi \psi''(\zeta)d\zeta d\xi \le \frac{1}{\nu_1}\int_1^t\int_1^\xi \psi''(\xi)\psi''(\zeta)d\zeta d\xi$$

$$= \frac{1}{\nu_1}\int_1^t \psi''(\xi)(\psi'(\xi) - \psi'(1))d\xi = \frac{1}{2\nu_1}\psi'(t)^2,$$

which gives (2.20).   □

Our next lemma presents some lower bounds for $|\psi'(t)|$ that can also be viewed as a characterization of the barrier behavior of the functions $\psi'(t)$ and $\psi(t)$.

Lemma 2.2.4 *Suppose that $\psi(t) \in \Omega_1$. Then*

$$|\psi'(t)| \ge \frac{\nu_1}{q}(t^{-q} - 1) \quad \forall t < 1$$

and

$$|\psi'(t)| \geq \frac{\nu_1}{p}(t^p - 1) \quad \forall t > 1.$$

Proof From SR1 we obtain, whenever $t < 1$,

$$\psi'(t) = \int_1^t \psi''(\xi)d\xi \leq \nu_1 \int_1^t (\xi^{p-1} + \xi^{-1-q})d\xi \leq \nu_1 \int_1^t \xi^{-1-q}d\xi = \frac{\nu_1}{q}(1 - t^{-q}).$$

This gives the first inequality of the lemma directly. One can prove the second inequality similarly.   □

The following lemma provides the means to bound the increase of $\psi(t)$ with respect to t.

Lemma 2.2.5 *Suppose that $\psi(t) \in \Omega_1$. Then, for any $\vartheta > 1$, we have*

$$\psi(\vartheta t) \leq \frac{\nu_2}{\nu_1}\left(\vartheta^{p+1}\psi(t) + \vartheta Y'_{p,q}(\vartheta)\sqrt{2\nu_1\psi(t)} + \nu_1 Y_{p,q}(\vartheta)\right).$$

Proof First we observe that if $\vartheta t \leq 1$ then $\psi(\vartheta t) < \psi(t)$, and then the inequality in the lemma is trivial, because ψ is strictly convex and minimal at $t = 1$. Thus, we may assume $\vartheta t > 1$. Now, if $t \leq 1$, then $1 < \vartheta t \leq \vartheta$, whence we have $\psi(\vartheta t) \leq \psi(\vartheta) \leq \nu_2 Y_{p,q}(\vartheta)$. Hence, if $t \leq 1$ the inequality in the lemma certainly holds. Therefore, from now on we assume that $t > 1$.
    According to (2.7), we have

$$\psi(\vartheta t) \leq \nu_2 Y_{p,q}(\vartheta t).$$

The rest of the proof consists of deriving an upper bound for $Y_{p,q}(\vartheta t)$. To simplify the notation we denote $Y_{p,q}$ by Y. From the definition of Y, we deduce that

$$Y(\vartheta t) = \frac{\vartheta^{p+1}t^{p+1} - 1}{p(p+1)} + \frac{\vartheta^{1-q}t^{1-q} - 1}{q(q-1)} + \frac{(p-q)(\vartheta t - 1)}{pq}.$$

By straightforward, although cumbersome, calculation, this can be rewritten as

$$Y(\vartheta t) = \vartheta^{1-q}\left(\frac{t^{p+1} - 1}{p(p+1)} + \frac{t^{1-q} - 1}{q(q-1)} + \frac{p-q}{pq}(t-1)\right) + \frac{(\vartheta^{p+1} - \vartheta^{1-q})}{p(p+1)}(t^{p+1} - 1)$$

$$+ \frac{(p-q)(\vartheta - \vartheta^{1-q})}{pq}(t-1) + \frac{\vartheta^{p+1} - 1}{p(p+1)} + \frac{\vartheta^{1-q} - 1}{q(q-1)} + \frac{p-q}{pq}(\vartheta - 1)$$

$$= \vartheta^{1-q}Y(t) + (\vartheta^{p+1} - \vartheta^{1-q})\frac{t^{p+1} - 1}{p(p+1)} + \frac{(p-q)(\vartheta - \vartheta^{1-q})}{pq}(t-1) + Y(\vartheta).$$

By using the second inequality (2.18) in Lemma 2.2.3, together with the fact that $\vartheta^{p+1} - \vartheta^{1-q} > 0$, we can conclude

$$Y(\vartheta t) \leq \vartheta^{1-q}Y(t) + (\vartheta^{p+1} - \vartheta^{1-q})\left(Y(t) + \frac{t-1}{p}\right)$$

$$+ \frac{(p-q)(\vartheta - \vartheta^{1-q})}{pq}(t-1) + Y(\vartheta)$$

$$= \vartheta^{p+1}Y(t) + \frac{\vartheta^{p+1} - \vartheta^{1-q}}{p}(t-1) + \frac{(p-q)(\vartheta - \vartheta^{1-q})}{pq}(t-1) + Y(\vartheta)$$

$$= \vartheta^{p+1}Y(t) + \vartheta\left(\frac{\vartheta^p - 1}{p} + \frac{1 - \vartheta^{-q}}{q}\right)(t-1) + Y(\vartheta)$$

$$= \vartheta^{p+1}Y(t) + \vartheta Y'(\vartheta)(t-1) + Y(\vartheta),$$

where the last equality follows from the choice of Y. Now, using the first inequality of (2.17) in Lemma 2.2.3, we obtain

$$Y(\vartheta t) \leq \vartheta^{p+1}Y(t) + \vartheta Y'(\vartheta)\sqrt{2Y(t)} + Y(\vartheta).$$

Finally, substitution of $Y(t) \leq \psi(t)/\nu_1$ yields the inequality in the lemma. $\square$

We remark that according to the above lemma, the growth behavior of the function $\psi(t)$ might be exponential with respect to the parameter p. An immediate consequence of Lemma 2.2.5 is the following.

Corollary 2.2.6 *Let $\psi(t) \in \Omega_1$ and $\vartheta > 1$. Then there exist two constants $\nu_3, \nu_4 > 0$ dependent on p and q such that*

$$\psi(\vartheta t) \leq \frac{\nu_2 \nu_4}{\nu_1}\left(\psi(t) + (\vartheta - 1)\sqrt{2\nu_1 \psi(t)} + \nu_1(\vartheta - 1)^2\right)$$

for all $\vartheta \in [1, 1 + \nu_3]$.

Proof With l'Hospital's rule we easily obtain the following limits:

$$\lim_{\vartheta \to 1} \frac{Y_{p,q}(\vartheta)}{(\vartheta - 1)^2} = 1, \quad \lim_{\vartheta \to 1} \frac{\vartheta Y'_{p,q}(\vartheta)}{\vartheta - 1} = 2.$$

These limits imply the existence of positive constants ν_3 and ν_4, such that for any $\vartheta \in [1, 1 + \nu_3]$,

$$\vartheta^{p+1} \leq \nu_4, \quad \vartheta \Upsilon'(\vartheta) \leq \nu_4(\vartheta - 1), \quad \Upsilon_{p,q}(\vartheta) \leq \nu_4(\vartheta - 1)^2.$$

Substitution of these inequalities into Lemma 2.2.5 yields the desired inequality.   □

2.3 RELATIONS BETWEEN S-R AND S-C FUNCTIONS

In this section, we discuss the relationship between the new class of self-regular functions and the class of so-called self-concordant functions in $\Re_{++}$. Let us first recall the definition of a univariate *self-concordant function* [83].

Definition 2.3.1   *A function* $\psi(t) \in C^3 : (0, \infty) \to \Re$ *is self-concordant if* ψ *is convex and satisfies the condition*

$$|\psi'''(t)| \leq \bar{\nu}(\psi''(t))^{3/2} \quad \forall t \in (0, \infty),$$

for some constant $\bar{\nu} > 0$.

In their remarkable book [83], Nesterov and Nemirovskii introduced the notion of *self-concordant barrier* to analyze IPMs for solving general convex optimization problems.

Definition 2.3.2   *A univariate function* $f(t) \in C^3$ *is said to be a self-concordant barrier for the domain* $\Re_+$ *if it satisfies the following two conditions:*

  (i)   *The function* $f(t)$ *is self-concordant.*
 (ii)   *There exists a constant* $\tilde{\nu}$ *such that*

$$|f'(t)| \leq \tilde{\nu}(f''(t))^{1/2} \quad \forall t \in (0, \infty).$$

An interesting property of a univariate *self-concordant* barrier is that, as observed by F. Glineur [29], there exists a constant $\hat{\nu}$ such that

$$f'(t)f'''(t) \leq \hat{\nu}(f''(t))^2.$$

This relation is precisely the conclusion presented in Proposition 2.2.2, where the function $\psi(t)$ is replaced by $f'(t)$. Since the set of self-concordant barriers is a subset of the set of self-concordant functions, in the sequel we focus only on the relations between the *self-concordant* functions and *self-regular* functions.
    We first observe that the self-regularity condition only requires that $\psi(t)$ be twice differentiable, whereas self-concordancy requires thrice differentiability. This does not mean, however, that the self-regularity condition is weaker than the condition for self-concordancy.
    By way of example, consider

$$\psi(t) = t - 1 - \log t,$$

which can be considered as the prototype of a univariate self-concordant function. Since

$$\psi'(t) = 1 - \frac{1}{t}, \quad \psi''(t) = \frac{1}{t^2}, \quad \psi'''(t) = -\frac{2}{t^3},$$

the self-concordancy of $\psi(t)$, with $\bar{\nu} = 2$, readily follows. It is easy to see that $\psi(t)$ does not satisfy condition SR1, since the second derivative $\psi''(t) = t^{-2}$ reduces to zero as t goes to infinity. Hence, $\psi(t)$ is not self-regular.

It is much harder to find a self-regular function that is not self-concordant. In fact, a large class of self-regular functions is self-concordant, as the next result shows. In this lemma we consider the functions

$$\psi(t) = \beta_0 \log t + \sum_{i=1}^{N} \beta_i(t^{\rho_i} - 1), \qquad (2.22)$$

introduced in Proposition 2.1.4.

Proposition 2.3.3 Let $\beta_0, \beta_1, \ldots \beta_N \in R$. If $\psi(t)$ in (2.22) belongs to Ω_1, then $\psi(t)$ is self-concordant.

Proof If $\psi(t) \in \Omega_1$, then let $\rho_o = 0$ and

$$p := \max\{\rho_i - 1 : i = 1, \ldots, N\} \geq 1, \quad q := \max\{1 - \rho_i : i = 0, 1, \ldots, N\} \geq 1.$$

The above definitions imply

$$t^{\rho_i - 2} \leq \max\{t^{p-1}, t^{-q-1}\} \quad \forall t > 0,$$

which allows us to write, for any $t > 0$,

$$t^{\rho_i - 3} \leq \max\{t^{3(\rho_i - 2)/2}, t^{-3}\} \leq \max\{t^{3(p-1)/2}, t^{-3}, t^{3(-q-1)/2}\}$$

$$= \max\{t^{3(p-1)/2}, t^{3(-q-1)/2}\} \leq (t^{p-1} + t^{-q-1})^{3/2},$$

where the first inequality follows from $t^{\rho_i} \leq \max\{t^{3\rho_i/2}, 1\}$. From the above inequality, one can conclude that there exists a constant $\bar{\nu}$ depending on the constants ν_1, ν_2 from (2.1), the coefficients β_i and the exponents ρ_i, such that

$$|\psi'''(t)| \leq \bar{\nu}(\psi''(t))^{3/2}.$$

This proves the proposition.   □

All examples of self-regular functions given so far were of the form (2.22). Thus we conclude that all self-regular functions considered so far are *self-concordant*.

We conclude this section by showing that there exist self-regular functions that are not *self-concordant*. To start with, we consider the function

$$\psi(t) = \int_1^t \int_1^\zeta \left(\xi^2 + \xi^{-2} + \frac{1}{2}\xi^2 \sin\xi + \frac{1}{2}\xi^{-2}\sin\xi^{-1} \right) d\xi d\zeta.$$

Note that $\psi(1) = \psi'(1) = 0$, and

$$\frac{1}{2}(t^2 + t^{-2}) \le \psi''(t) = t^2 + t^{-2} + \frac{1}{2}t^2 \sin t + \frac{1}{2}t^{-2}\sin t^{-1} \le \frac{3}{2}(t^2 + t^{-2}),$$

proving that $\psi(t) \in \Omega_1$. Now we show that this function is not self-concordant. One has

$$\psi'''(t) = 2t - 2t^{-3} + \frac{1}{2}t^2 \cos t + t \sin t - t^{-3}\sin t^{-1} - \frac{1}{2}t^{-4}\cos t^{-1}.$$

Now consider a sequence t_k of positive numbers converging to 0, and such that $\cos t_k^{-1} = -1$ for each k. Then, for each k,

$$\psi'''(t_k) = 2t_k - 2t_k^{-3} + \frac{1}{2}t_k^2 \cos t_k + t_k \sin t_k + \frac{1}{2}t_k^{-4}.$$

Clearly, if k goes to infinity and hence t_k decreases to 0, then the behavior of $\psi'''(t_k)$ is dominated by the last term, $1/2t_k^{-4}$. As a consequence, the quotient

$$\frac{|\psi'''(t_k)|}{(\psi''(t_k))^{3/2}}$$

is unbounded when t_k goes to 0, since the behavior of the denominator is then dominated by a term of order t_k^{-3}. This proves that $\psi(t)$ is not self-concordant. Unfortunately, we were not able to show that $\psi(t)$ belongs to Ω_2. We conclude with another example showing that the class of self-regular function is not contained in the class of self-concordant functions.

Example 2.3.4 We consider the function

$$\psi(t) = (t - t^{-1})^2 + \frac{1}{3}\left(\int_1^t \sin \zeta^{-3} d\zeta + \int_1^{t^{-1}} \sin \zeta^{-3} d\zeta \right)$$

$$= (t - t^{-1})^2 + \frac{1}{3}\int_1^t (\sin \zeta^{-3} - \zeta^{-2}\sin \zeta^3)d\zeta.$$

Obviously, $\psi(1) = \psi'(1) = 0$. By direct calculus, we get

$$\psi'(t) = 2t - 2t^{-3} + \frac{1}{3}\sin t^{-3} - \frac{1}{3}t^{-2}\sin t^3$$

$$\psi''(t) = 2 + 6t^{-4} - t^{-4}\cos t^{-3} + \frac{2}{3}t^{-3}\sin t^3 - \cos t^3.$$

Since for any $t > 0$,

$$-1 \le \cos t^{-3}, \sin t^3, \cos t^3 \le 1, \quad t^{-4} \ge \frac{4}{3}t^{-3} - \frac{1}{3},$$

it follows that

$$\psi''(t) \ge 1 + 5t^{-4} - \frac{2}{3}t^{-3} \ge \frac{5}{6} + \frac{9}{2}t^{-4} > \frac{5}{6}(1 + t^{-4})$$

and

$$\psi''(t) \le 3 + 7t^{-4} + \frac{2}{3}t^{-3} < \frac{15}{2}(1 + t^{-4}).$$

These two inequalities imply $\psi(t) \in \Omega_1$. Since $\psi(t) = \psi(t^{-1})$ for any $t > 0$, Lemma 2.1.8 implies that $\psi(t)$ is in Ω_2. Hence $\psi(t)$ is *self-regular*.

We proceed by showing that $\psi(t)$ is not self-concordant. One has

$$\psi'''(t) = -24t^{-5} - 4t^{-5}\cos t^{-3} - 3t^{-8}\sin t^{-3} - 2t^{-4}\sin t^3 + 2t^{-1}\cos t^3 + 3t^2\sin t^3.$$

As in the previous example, we consider a sequence t_k of positive numbers converging to 0, such that $\sin t_k^{-3} = -1$ for each k. Then, if k goes to infinity and thus t_k decreases to 0, the behavior of $\psi'''(t_k)$ is dominated by the term of order t_k^{-8}. As a consequence, the quotient

$$\frac{|\psi'''(t_k)|}{(\psi''(t_k))^{3/2}}$$

is unbounded if t_k goes to 0, since the behavior of the denominator is then dominated by a term of order t_k^{-6}. This proves that $\psi(t)$ is not self-concordant.

Figure 2.3 illustrates the relations and differences between *self-regular* and *self-concordant* functions.

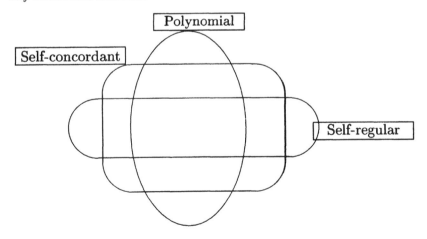

Figure 2.3 Relations among Self-regular, Self-concordant and Polynomial functions.

From the results of this section it is clear that the set of self-regular functions and the set of self-concordant functions are different, but there is really a large intersection of these two sets. The functions in the intersection share all the good properties of self-concordant functions, but as we will show, the additional property of being self-regular enables us to get a better complexity result for large-update IPMs than can be obtained on the basis of self-concordancy alone.

Chapter 3

Primal-Dual Algorithms for Linear Optimization Based on Self-Regular Proximities

The notion of self-regular functions *is extended to the positive orthant* $\Re_{++}^n$. *Then, self-regular proximities of LO are introduced based on self-regular functions and the so-called v-space. Several attractive features of self-regular proximity are explored. A new class of primal-dual Newton-type methods relying on these self-regular proximities for solving LO problems is proposed and the complexity results of large-update and small-update algorithms are established. The procedure of complexity analysis in this chapter sets up a pattern for the complexity analysis in later chapters. The possibility of relaxing the* self-regularity *requirement on the proximity used in the algorithm for solving LO is also discussed.*

3.1 SELF-REGULAR FUNCTIONS IN $\mathfrak{R}_{++}^n$ AND SELF-REGULAR PROXIMITIES FOR LO

We start with the definition of self-regular functions on $\mathfrak{R}_{++}^n$. For simplicity, if no confusion occurs, sometimes we abuse some of the notation such as the function $\psi(\cdot)$.

Definition 3.1.1   *A function* $\psi(x) : \mathfrak{R}_{++}^n \longrightarrow \mathfrak{R}_+^n$ *defined by*

$$\psi(x) = (\psi(x_1), \dots, \psi(x_n))^{\mathrm{T}}, \tag{3.1}$$

is said to be self-regular if the kernel function $\psi(t) : \mathfrak{R}_{++} \longrightarrow \mathfrak{R}_+$ *is self-regular.*

Note that from (3.1), we can define the power x^r for any real number r when $x > 0$, $\Psi'(x) \in \mathfrak{R}^n$ and $\Psi''(x) \in \mathfrak{R}^n$.

To maintain consistency of notation in this context, we denote by $\Psi(x)$: $\mathfrak{R}_{++}^n \longrightarrow \mathfrak{R}_+$ the function

$$\Psi(x) := \sum_{i=1}^n \psi(x_i). \tag{3.2}$$

From definitions (3.1) and (3.2), one can easily see that self-regular functions in $\mathfrak{R}_{++}^n$ also enjoy certain attractive features inherited directly from their kernel functions. These interesting characterizations of self-regular functions in the positive orthant are collected in our next proposition. To avoid repetition, we omit the detailed proof of the proposition.

Proposition 3.1.2   *Let the functions* $\psi(x)$ *and* $\Psi(x)$ *be defined by (3.1) and (3.2). If the kernel function* $\psi(t)$ *is self-regular with parameters* $\nu_2 \geq \nu_1 > 0$ *and* $p, q \geq 1$, *then*

(i) $\Psi(x)$ *is strictly convex with respect to* $x \in \mathfrak{R}_{++}^n$, *and vanishes at its global minimal point, the all-one vector* $x = e$, *that is,* $\Psi(e) = 0$ *and* $\psi(e) = \psi'(e) = 0$. *Further,*

$$\nu_1(x^{p-1} + x^{-1-q}) \leq \psi''(x) \leq \nu_2(x^{p-1} + x^{-1-q}) \quad \forall x \in \mathfrak{R}_{++}^n; \tag{3.3}$$

(ii)   *For any* $x, s \in \mathfrak{R}_{++}^n$,

$$\psi(x^r s^{1-r}) \leq r\psi(x) + (1-r)\psi(s) \quad \forall r \in [0, 1]. \tag{3.4}$$

We now consider the behavior of $\Psi(x)$ when $x(t) = (x_1(t), \dots, x_n(t))^{\mathrm{T}}$ is a function of t. For notational convenience, we denote by $x'(t), x''(t)$ the first and second derivatives of $x(t)$ with respect to t, separately, that is,

$$x'(t) = (x_1'(t), \dots, x_n'(t))^{\mathrm{T}}, \quad x''(t) = (x_1''(t), \dots, x_n''(t))^{\mathrm{T}}. \tag{3.5}$$

Now, with simple calculus, one can easily prove the following results, which further provide tools to estimate the first and second derivatives of $\Psi(x(t))$ with respect to t.

Lemma 3.1.3 *Suppose that $x(t)$ is a function such that the vector $x(t) \in \Re_{++}^n$. If $x(t)$ is twice differentiable with respect to t for all $t \in (l_t, u_t)$ and $\psi(t)$ is also a twice continuously differentiable function in a suitable domain that contains all the components of $x(t)$, then*

$$\frac{d}{dt} \Psi(x(t)) = \psi'(x(t))^{\mathrm{T}} x'(t), \quad t \in (l_t, u_t),$$

$$\frac{d^2}{dt^2} \Psi(x(t)) \leq \varpi \|x'(t)\|^2 + \psi'(x(t))^{\mathrm{T}} x''(t), \tag{3.6}$$

where

$$\varpi = \max\{|\psi''(x_i(t))|, \ i = 1, 2, \ldots, n\} \tag{3.7}$$

is a number depending on $x(t)$ and $\psi(t)$.

Now let us move onto the main theme in this work, the new proximity and search direction based on it. Recalling the notation introduced in Sections 1.2.3 and 1.3.2, we can rewrite the original Newton system for LO problem (1.6) as a new Newton system (1.14) in the scaled v-space. This scaling technique is widely applied in the IPM literature to ease the notation and analysis of IPMs.

The proximity we suggest is defined as follows:

$$\Phi(x, s, \mu) := \Psi(v) = \sum_{i=1}^n \psi(v_i). \tag{3.8}$$

We say the proximity is *self-regular* if the function $\psi(t)$ is *self-regular*. Recall that, as described in the introduction, the targeted center in the primal and dual space is denoted by $x(\mu), s(\mu)$, respectively. In consequence, we can define the functions $\Psi(x(\mu)^{-1}x)$ and $\Psi(s(\mu)^{-1}s)$ as proximities in the primal and dual space. In the sequel we give a relation between the proximities in the v-space, the primal and the dual space.

Proposition 3.1.4 *Let the proximity $\Psi(v)$ be defined by (3.8). If it is self-regular, then*

$$\Psi(v) \leq \frac{1}{2}\left(\Psi\left(x(\mu)^{-1}x\right) + \Psi\left(s(\mu)^{-1}s\right)\right).$$

Proof The conclusion follows readily from the fact that

$$v_i = (x(\mu)^{-1/2}x_i^{1/2})(s(\mu)^{-1/2}s_i^{1/2}) \quad \forall i = 1, 2, ..., n,$$

and condition SR2. $\square$

The above proposition provides us with a mean to bound the proximity measure in the scaled v-space according to the values of its counterparts in the primal "x" and dual "s" spaces. The proposition is simply based on condition SR2. As we see later, this condition substantially simplifies the analysis of the decreasing behavior of the proximity measure.

Our next result summarizes some appealing properties of the proximity $\Psi(v)$ that are also shared by general self-regular functions in $\Re^n_{++}$. For notational convenience, we also define

$$\sigma = \|\nabla\Psi(v)\|. \tag{3.9}$$

Proposition 3.1.5 *Let the proximity $\Psi(v)$ be defined by (3.8). Then*

$$\Psi(v) \leq \frac{\sigma^2}{2\nu_1}, \tag{3.10}$$

$$v_{\min} \geq \left(1 + \frac{q\sigma}{\nu_2}\right)^{-\frac{1}{q}}, \tag{3.11}$$

and

$$v_{\max} \leq \left(1 + \frac{p\sigma}{\nu_1}\right)^{\frac{1}{p}}. \tag{3.12}$$

If $v_{\max} > 1$ and $v_{\min} < 1$, then

$$\sigma \geq \nu_1 \left(\frac{(v^p_{\max} - 1)^2}{p^2} + \frac{(v^{-q}_{\min} - 1)^2}{q^2}\right)^{\frac{1}{2}}. \tag{3.13}$$

For any $\vartheta > 1$,

$$\Psi(\vartheta v) \leq \frac{\nu_2}{\nu_1}\left(\vartheta^{p+1}\Psi(v) + \vartheta\Upsilon'_{p,q}(\vartheta)\sqrt{2n\nu_1\Psi(v)} + n\nu_1\Upsilon_{p,q}(\vartheta)\right). \tag{3.14}$$

If $\vartheta \in (1, 1 + \nu_3]$, then

$$\Psi(\vartheta v) \leq \frac{\nu_2\nu_4}{\nu_1}\Psi(v) + \frac{\nu_2\nu_4\sqrt{2n\nu_1\Psi(v)}}{\nu_1}(\vartheta - 1) + n\nu_2\nu_4(\vartheta - 1)^2, \tag{3.15}$$

where ν_3, ν_4 are the same constants as introduced in Corollary 2.2.6.

Proof The first inequality of the proposition follows from (2.20) in Lemma 2.2.3. To prove the second inequality, we first note that it holds trivially if $v_{\min} \geq 1$. If $v_{\min} < 1$, from Lemma 2.2.4 one obtains

$$\sigma \geq |\psi'(v_{\min})| \geq \frac{\nu_2}{q}(v_{\min}^{-q} - 1),$$

which implies (3.11). By following an analogous process one can show that both (3.12) and (3.13) are true. Inequality (3.14) follows immediately from Lemma 2.2.5, if we apply the Cauchy–Schwarz inequality to the all-one vector e and the vector $(\sqrt{\psi(v_1)}, \ldots, \sqrt{\psi(v_n)})^{\mathrm{T}}$. By using (3.14) and following a similar chain of reasoning as in the proof of Corollary 2.2.6, one can readily obtain the last conclusion of the proposition. $\square$

Inequality (3.15) implies that, if $\Psi(v) \leq \tau$ with $\tau > 0$, and $\vartheta - 1 = \nu_3/\sqrt{n} \leq \nu_3$, then

$$\Psi(\vartheta v) \leq \frac{\nu_2 \nu_4 \tau}{\nu_1} + \nu_2 \nu_3 \nu_4 \sqrt{\frac{2\tau}{\nu_1}} + \nu_2 \nu_4 \nu_3^2. \qquad (3.16)$$

Figure 3.1 exhibits some neighborhoods defined by *self-regular* proximities. The picture is drawn in v-space with all level sets given by $\Psi(v) \leq 3$. Observe that the neighborhood shrinks as p and q increase.

It is instructive to discuss the relations between the duality gap and the new proximity. Note that from Lemma 2.2.3 we readily obtain

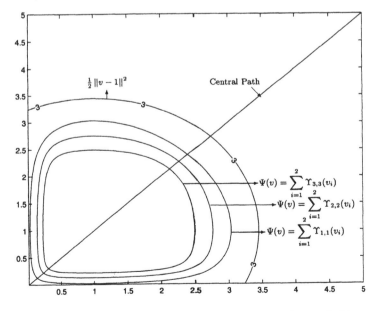

Figure 3.1   Various large neighborhoods defined by *self-regular* proximities.

$$\frac{\Psi(v)}{\nu_1} \geq \frac{1}{2}\|v - e\|^2 = \frac{1}{2}\|v\|^2 - \sum_{i=1}^n v_i + \frac{n}{2} \geq \frac{1}{2}\|v\|^2 - \sqrt{n}\|v\| + \frac{n}{2},$$

which further implies

$$\|v\| \leq \sqrt{n} + \sqrt{\frac{2\Psi(v)}{\nu_1}}.$$

It follows immediately that

$$x^T s = \mu\|v\|^2 \leq n\mu + 2\mu\sqrt{2\frac{n\Psi(v)}{\nu_1}} + \frac{2\Psi(v)}{\nu_1}\mu. \tag{3.17}$$

This implies that the proximity can be used as a potential function for minimizing the duality gap.

It is worth thinking a little more about which kind of *self-concordant* functions are not *self-regular*. Let us recall the example $\psi(t) = t - 1 - \log t$. This function is *self-concordant* but not *self-regular*. To be more specific, let us consider the corresponding proximity $\Psi(v) = \sum_{i=1}^n (v_i - 1 - \log v_i)$. Our target is to minimize the duality gap $x^T s$ while staying in a suitable neighborhood of the central path. When the parameter μ is fixed, this is equivalent to minimizing the argument $\|v\|^2$ in v-space. Naturally we expect the proximity measure to act as a potential function for the scaled duality gap $\|v^2\|$ or the square of the distance to the central path $\|v - e\|^2$, or in other words, $\Psi(v) \geq \hat{c}\max(\|v\|^2, \|v - e\|^2)$ for some constant $\hat{c} > 0$. However, with a closer look one can easily see that such a constant $\hat{c}$ does not exist for all $v \in \Re^n_{++}$. Hence, the proximity measure is not "good". Thus we consider this function not suitable for our purpose.

Similarly, the function $\psi(t) = t^2 - 2t + 1$ is also *self-concordant* but not *self-regular*, since there is no barrier term in it. One can see that the corresponding proximity $\Psi(v) = \|v - e\|^2$ does not have good control on the distance to the boundary of the feasible set, since when $\|v\|$ goes to zero, the proximity measure $\Psi(v)$ does not increase to infinity. Again we exclude this proximity from our "regular" group.

3.2 THE ALGORITHM

Now we are ready to introduce the new search direction for solving LO problems. Let $d_x, \Delta y, d_s$ denote the solution of the following modified Newton equation system for the scaled system 1.13:

$$\overline{A}d_x = 0,$$

$$\overline{A}^T \Delta y + d_s = 0,$$

$$d_x + d_s = -\nabla\Psi(v),$$

or equivalently, let $(\Delta x, \Delta y, \Delta s)$ solve the system

$$A\Delta x = 0,$$

$$A^T\Delta y + \Delta s = 0, \tag{3.18}$$

$$s\Delta x + x\Delta s = -\sqrt{\mu xs}\ \nabla\Psi\left(\sqrt{xs/\mu}\right).$$

The last equation in (3.18) represents the Newton system for the equation

$$xs = -\mu v(\nabla\Psi(v) - v),$$

where the right-hand side is fixed. We would like to remark that when we use the new proximity $\Psi(v)$ with a large value q to define the neighborhood and then employ the new system (3.18), the small entries $x_i s_i$ (or equivalently v_i) are heavily penalized in terms of the neighborhood as well as in the definition of the search direction.

Recall (cf. chapter 7 in [98]) that if rank$(A) = m$, then for any $\mu > 0$, system (3.18) has a unique solution $\Delta x, \Delta y, \Delta s$. The result of a damped Newton step with damping factor α is denoted by

$$x_+ = x + \alpha\Delta x, \quad y_+ = y + \alpha\Delta y, \quad s_+ = s + \alpha\Delta s. \tag{3.19}$$

In the algorithm we use a threshold value τ for the proximity and we assume that we are given a triple (x^0, y^0, s^0) such that $\Phi(x^0, s^0, \mu^0) \le \tau$ for $\mu^0 = 1$. This can be done without loss of generality (cf. [98]).

If for the current iterates (x, y, s) and barrier parameter value μ, the proximity $\Psi(x, s, \mu)$ exceeds τ, then we use one or more damped Newton steps to recenter; otherwise μ is reduced by the factor $1 - \theta$. This is repeated until $n\mu < \varepsilon$. Thus the algorithm can be stated as follows:

Primal-Dual Algorithm for LO

Inputs
A proximity parameter $\tau > \nu_1^{-1}$;
an accuracy parameter $\varepsilon > 0$;
a fixed barrier update parameter $\theta, 0 < \theta < 1$;
(x^0, s^0) and $\mu^0 = 1$ such that $\Phi(x^0, s^0, \mu^0) \le \tau$.

begin
  $x := x^0$; $s := s^0$; $\mu := \mu^0$;
  while $n\mu \ge \varepsilon$ **do**
  begin
    $\mu := (1 - \theta)\mu$;
    while $\Phi(x, s, \mu) \ge \tau$ **do**

begin
    Solve system (3.18) for Δx, Δy, Δs;
    Determine a step size α;
    $x := x + \alpha \Delta x$;
    $y := y + \alpha \Delta y$;
    $s := s + \alpha \Delta s$;
  end
 end
end

Remark 3.2.1 *At each inner iteration, the step size α has to be chosen such that the proximity Ψ decreases sufficiently. In the next section we present a default value for α, based on the value of the proximity.*

Remark 3.2.2 *There are various choices for the parameter τ. If τ is a small constant independent of n and $\theta = \mathcal{O}(1/\sqrt{n})$, then the algorithm is called an IPM with small neighborhood, which has the best known $\mathcal{O}(\sqrt{n} \log n/\varepsilon)$ iteration bound. If τ is chosen to be a number related to n, for instance $\tau = n$, then we call the algorithm an IPM with large neighborhood. For large neighborhood IPMs, the current best known iteration bound, based on a self-concordant function, is $\mathcal{O}(n \log n/\varepsilon)$. This is the bound we want to improve by using a self-regular proximity measure.*

Remark 3.2.3 *In the algorithm we always assume that $v_{\max} > 1$. This is because when $v_{\max} \leq 1$, we can reduce the value of the proximity in the algorithm (or stay in a certain neighborhood of the central path) by appropriately reducing μ. In such a case we do not even need to solve the Newton-type system.*

Remark 3.2.4 *The algorithm terminates with a point satisfying $n\mu \leq \varepsilon$ and $\Phi(x, s, \mu) \leq \tau$. By using (3.17), we obtain*

$$x^{\mathrm{T}} s \leq n\mu + 2\mu \sqrt{\frac{2n\tau}{\nu_1}} + \mu \frac{2\tau}{\nu_1}.$$

Hence, if $\tau = \mathcal{O}(n)$, which means that the algorithm works indeed in a large neighborhood of the central path, then the algorithm finally reports a feasible solution such that $x^{\mathrm{T}} s = \mathcal{O}(\varepsilon)$. For instance, if we choose the parameter $\tau = n$ and the proximity satisfying condition SR1 with $\nu_1 = 1$, then the algorithm will provide a solution satisfying $x^{\mathrm{T}} s \leq 7\varepsilon$.

3.3 ESTIMATE OF THE PROXIMITY AFTER A NEWTON STEP

This section is devoted to investigate the decrease of the proximity after a strictly feasible step. Note that once the solution of the system (3.18), or equivalently the search direction, is available, we usually want to know how far we can go along the search direction while staying in the feasible region. As we mentioned in Section 1.2.3 (see also [7]), in most successful IPM solvers, the step size is chosen as a fixed factor of the maximal step size. Hence, the evaluation of the maximal feasible step size is very helpful for both the theoretical analysis and efficient implementation of IPMs. A useful observation is that, since the displacements Δx and Δs are orthogonal, the scaled displacements d_x and d_s are orthogonal as well, that is,

$$d_x^{\mathrm{T}} d_s = 0, \tag{3.20}$$

which further yields, by (3.9),

$$\sigma^2 = \| - \nabla\Psi(v)\|^2 = (d_x + d_s)^{\mathrm{T}}(d_x + d_s) = \|d_x\|^2 + \|d_s\|^2. \tag{3.21}$$

Let us define

$$v_+ = \sqrt{\frac{x_+ s_+}{\mu}} = \sqrt{(v + \alpha d_x)(v + \alpha d_s)},$$

where x_+, s_+ are defined in (3.19). It is easy to see that x_+ and s_+ are feasible if and only if $v + \alpha d_x$ and $v + \alpha d_s$ are both nonnegative. For simplicity, we define

$$\bar{d}_x := \frac{\Delta x}{x}, \quad \bar{d}_s := \frac{\Delta s}{s}. \tag{3.22}$$

It follows that

$$v + \alpha d_x = v(e + \alpha \bar{d}_x), \quad v + \alpha d_s = v(e + \alpha \bar{d}_s).$$

This further implies that the maximal step size $\alpha_{\max}$ is essentially determined by the conditions

$$e + \alpha_{\max} \bar{d}_x \geq 0, \quad e + \alpha_{\max} \bar{d}_s \geq 0.$$

Our first result in this section provides an estimate of the norm of $(\bar{d}_x, \bar{d}_s)$. Such estimation further gives a bound for $\alpha_{\max}$, because the maximal feasible step size $\alpha_{\max}$ certainly satisfies $\alpha_{\max} \|(\bar{d}_x, \bar{d}_s)\| \geq 1$.

Lemma 3.3.1 *Let $\bar{d}_x, \bar{d}_s$ be defined by (3.22) and let*

$$\bar{\alpha} = v_{\min} \sigma^{-1}. \tag{3.23}$$

Then

$$\|(\bar{d}_x, \bar{d}_s)\| \leq \bar{\alpha}^{-1},$$

and the maximal feasible step size $\alpha_{\max}$ *satisfies*

$$\alpha_{\max} \geq \bar{\alpha} \geq \sigma^{-1}\left(1 + \frac{q\sigma}{\nu_1}\right)^{-1/q}.$$

Proof   It follows directly from (3.21)

$$\|(d_x, d_s)\|^2 = \sigma^2.$$

By using the notation $(\bar{d}_x, \bar{d}_s)$ (3.22), we obtain

$$\|(\bar{d}_x, \bar{d}_s)\| = \|(v^{-1}d_x, v^{-1}d_s)\| \leq \frac{1}{v_{\min}}\|(d_x, d_s)\| \leq \frac{\sigma}{v_{\min}} = \sigma\left(\frac{1 + q\sigma}{\nu_1}\right)^{1/q},$$

where the last inequality is derived from (3.11). This gives the first inequality in the lemma, from which the second inequality can be easily derived. $\square$

We next proceed to estimate the decrease of the proximity value after one step. It is apparent that the proximity after one step is given by $\Psi(v_+)$. Let us denote the change in the proximity before and after one step as a function of α, that is,

$$g(\alpha) = \Psi(v_+) - \Psi(v). \tag{3.24}$$

From the definitions of $\psi(x)$ by (3.1) and $\Psi(x)$ by (3.2), and by making use of the self-regularity of $\psi(x)$, we derive

$$g(\alpha) \leq -\Psi(v) + \frac{1}{2}\sum_{i=1}^{n}(\psi(v_i + \alpha[d_x]_i) + \psi(v_i + \alpha[d_s]_i)). \tag{3.25}$$

Let us denote by $g_1(\alpha)$ the expression on the right-hand side of (3.25). Obviously both the functions $g(\alpha)$ and $g_1(\alpha)$ are twice continuously differentiable with respect to α if the step size α is feasible. To evaluate the decrease of the function $g(\alpha)$ after one feasible step, it suffices to estimate the value of the function $g_1(\alpha)$ for a feasible step size α. An important step in estimating $g_1(\alpha)$ is to bound its second derivative, which is done in our next lemma.

Lemma 3.3.2 *Let the function $g_1(\alpha)$ be defined by the right-hand side of (3.25) and assume that the parameter $\alpha \in [0, \bar{\alpha})$. Then*

$$g_1''(\alpha) \leq \frac{\nu_2\sigma^2}{2}\left((v_{\max} + \alpha\sigma)^{p-1} + (v_{\min} - \alpha\sigma)^{-q-1}\right). \tag{3.26}$$

Proof   From Lemma 3.1.3 and (3.25) it follows that

$$g_1''(\alpha) \leq \frac{1}{2} \max\{\psi''(v_i + \alpha[d_x]_i) : i \in I\}\|d_x\|^2$$

$$+ \frac{1}{2} \max\{\psi''(v_i + \alpha[d_s]_i) : i \in I\}\|d_s\|^2$$

$$\leq \frac{\sigma^2}{2} \max\{\psi''(v_i + \alpha[d_x]_i), \psi''(v_i + \alpha[d_x]_i) : i \in I\}$$

$$\leq \frac{\nu_2 \sigma^2}{2} ((v_{\max} + \alpha\sigma)^{p-1} + (v_{\min} - \alpha\sigma)^{-1-q}),$$

where the second inequality follows from the definition of σ, and the third inequality is implied by condition SR1 and the fact that for any $i \in I$,

$$v_{\min} - \alpha\sigma \leq v_{\min} - \alpha\|d_x\| \leq v_i + \alpha[d_x]_i \leq v_{\max} + \alpha\|d_x\| \leq v_{\max} + \alpha\sigma,$$
(3.27)

$$v_{\min} - \alpha\sigma \leq v_{\min} - \alpha\|d_s\| \leq v_i + \alpha[d_s]_i \leq v_{\max} + \alpha\|d_s\| \leq v_{\max} + \alpha\sigma.$$
(3.28)

The proof of the lemma is completed. $\qquad\square$

By means of direct calculus and applying Lemma 3.1.3, one gets

$$g(0) = g_1(0) = 0, \quad g_1'(0) = g'(0) = -\frac{\sigma^2}{2}.$$

Now, using $g_1(\alpha) = g_1(0) + g_1'(0)\alpha + \int_0^\alpha \int_0^\xi g_1''(\zeta)d\zeta d\xi$ and (3.26), we conclude that for any $\alpha \in [0, \bar{\alpha})$,

$$g_1(\alpha) \leq -\frac{\sigma^2}{2}\alpha + \frac{\nu_2 \sigma^2}{2} \int_0^\alpha \int_0^\xi \left((v_{\max} + \zeta\sigma)^{p-1} + (v_{\min} - \zeta\sigma)^{-q-1}\right)d\zeta d\xi.$$
(3.29)

Denote by $g_2(\alpha)$ the expression on the right-hand side of (3.29). Clearly $g_2(\alpha)$ is convex and twice differentiable in the interval $[0, \bar{\alpha})$. Let us denote by α^* the global minimizer of $g_2(\alpha)$ in the interval $[0, \bar{\alpha})$, that is

$$\alpha^* = \arg \min_{\alpha \in [0, \bar{\alpha})} g_2(\alpha).$$
(3.30)

There is no difficulty in checking that $g_2'(0) < 0$ and $g_2'(\alpha)$ goes to infinity as α approaches $\bar{\alpha}$. From the convexity of $g_2(\alpha)$ in $[0, \bar{\alpha})$ one can easily deduce that α^* is the unique solution of the equation

$$-\sigma + \frac{\nu_2}{p}((v_{\max} + \alpha^*\sigma)^p - v_{\max}^p) + \frac{\nu_2}{q}((v_{\min} - \alpha^*\sigma)^{-q} - v_{\min}^{-q}) = 0. \quad (3.31)$$

Thus, one has

$$\alpha^* \leq \bar{\alpha}.$$

In what follows we present a lower bound on the value of α^*.

Lemma 3.3.3 *Let the constant* α^* *be defined by (3.30). Suppose that* $\Psi(v) \geq \nu_1^{-1}$ *and* $v_{\max} > 1$. *Let*

$$\nu_5 := \min\left\{\frac{\nu_1}{2\nu_2(p+\nu_1)+\nu_1(p-1)}, \frac{\nu_1^2}{(1+\nu_1)(2\nu_1\nu_2+q(\nu_1+2\nu_2))}\right\}.$$

Then we have

$$\alpha^* \geq \nu_5 \sigma^{-(q+1)/q}. \tag{3.32}$$

In the special case where $\psi(t) = Y_{p,q}(t)$ *given by (2.5), the above bound simplifies to*

$$\alpha^* \geq \min\left\{\frac{1}{3p+1}, \frac{1}{6q+4}\right\}\sigma^{-(q+1)/q}. \tag{3.33}$$

Proof Since $\Psi(v) \geq \nu_1^{-1}$, by Proposition 3.1.5 we have $\sigma \geq 1$. Let us define

$$\omega_1(\alpha) = -\frac{\sigma}{2} + \frac{\nu_2}{p}\left((v_{\max} + \alpha\sigma)^p - v_{\max}^p\right)$$

and

$$\omega_2(\alpha) = -\frac{\sigma}{2} + \frac{\nu_2}{q}\left((v_{\min} - \alpha\sigma)^{-q} - v_{\min}^{-q}\right).$$

It is easy to verify that both functions $\omega_1(\alpha)$ and $\omega_2(\alpha)$ are increasing for $\alpha \in [0, \bar{\alpha})$. Using these two functions, we can rewrite equation (3.31) as

$$\omega_1(\alpha^*) + \omega_2(\alpha^*) = 0.$$

Through simple calculus, we know that $\omega_1(\alpha_1^*) = 0$ if

$$\alpha_1^* = \frac{v_{\max}}{\sigma}\left(\left(1 + \frac{p\sigma}{2\nu_2 v_{\max}^p}\right)^{1/p} - 1\right).$$

We now progress to estimate α_1^*. First observe that, since $p \geq 1$, by using (1.10) in Lemma 1.3.1, we obtain

$$\left(1 + \frac{p\sigma}{2\nu_2 v_{\max}^p}\right)^{1/p} = \left(1 - \frac{p\sigma}{p\sigma + 2\nu_2 v_{\max}^p}\right)^{-1/p}$$

$$\geq \left(1 - \frac{\sigma}{p\sigma + 2\nu_2 v_{\max}^p}\right)^{-1} = 1 + \frac{\sigma}{(p-1)\sigma + 2\nu_2 v_{\max}^p}.$$

It follows readily that

$$\alpha_1^* \geq \frac{v_{\max}}{(p-1)\sigma + 2v_2 v_{\max}^p} \geq \frac{1}{(p-1)\sigma + 2v_2 v_{\max}^p}, \tag{3.34}$$

where the last inequality is given by the assumption that $v_{\max} \geq 1$. Because $p \geq 1$, from (3.12) we can conclude that

$$v_{\max}^p \leq \frac{p\sigma}{v_1} + 1 \leq \frac{p\sigma}{v_1} + \sigma.$$

This relation, along with (3.34), yields

$$\alpha_1^* \geq \frac{v_1}{2v_2(p + v_1) + v_1(p-1)} \sigma^{-1}.$$

We next estimate the root of $w_2(\alpha)$, or in other words the value

$$\alpha_2^* = \frac{v_{\min}}{\sigma} \left(1 - \left(1 + \frac{q\sigma v_{\min}^q}{2v_2} \right)^{-1/q} \right).$$

Again, by applying (1.10), one gets

$$\left(1 + \frac{q\sigma v_{\min}^q}{2v_2} \right)^{-1/q} = \left(1 - \frac{q\sigma v_{\min}^q}{2v_2 + q\sigma v_{\min}^q} \right)^{1/q} \leq 1 - \frac{\sigma v_{\min}^q}{2v_2 + q\sigma v_{\min}^q}.$$

Hence

$$\alpha_2^* \geq \frac{v_{\min}}{\sigma} \frac{\sigma v_{\min}^q}{2v_2 + q\sigma v_{\min}^q} = \frac{v_{\min}^{q+1}}{2v_2 + q\sigma v_{\min}^q}. \tag{3.35}$$

Recalling (3.11) one obtains

$$v_{\min}^q \geq \frac{v_1}{v_1 + q\sigma}, \tag{3.36}$$

which, together with (3.35), further implies

$$\alpha_2^* \geq \frac{v_1 v_{\min}}{2v_1 v_2 + q(v_1 + 2v_2)} \sigma^{-1}. \tag{3.37}$$

Since $\sigma \geq 1$, from (3.36) it immediately follows that

$$v_{\min} \geq \left(1 + \frac{q\sigma}{v_1} \right)^{-1/q} \geq \left(1 + \frac{q}{v_1} \right)^{-1/q} \sigma^{-1/q} \geq \frac{v_1}{1 + v_1} \sigma^{-1/q}. \tag{3.38}$$

where the last inequality is true because, by using (1.10) in Lemma 1.3.1, we have

$$\left(1 + \frac{q}{v_1} \right)^{1/q} \leq 1 + \frac{1}{v_1}.$$

Combining (3.37) with (3.38), one can conclude that

$$\alpha_2^* \geq \frac{\nu_1^2}{(1 + \nu_1)(2\nu_1\nu_2 + q(\nu_1 + 2\nu_2))} \sigma^{-(q+1)/q}.$$

Now let us recall the fact that both $w_1(\alpha)$ and $w_2(\alpha)$ are increasing functions of α. From the above discussions we conclude that

$$\alpha^* \geq \min\{a_1^*, a_2^*\} \geq \nu_5 \sigma^{-(q+1)/q},$$

where

$$\nu_5 = \min\left\{\frac{\nu_1}{2\nu_2(p + \nu_1) + \nu_1(p - 1)}, \frac{\nu_1^2}{(1 + \nu_1)(2\nu_1\nu_2 + q(\nu_1 + 2\nu_2))}\right\}$$

is a constant depending only on the proximity $\Psi(v)$ and independent of the underlying problem. This proves the first conclusion of the lemma. The second statement in the lemma is a direct consequence of inequality (3.32). □

It is straightforward to verify that the function $g_2(\alpha)$ satisfies all the conditions in Lemma 1.3.3. Therefore, at the point α^* defined by (3.30),

$$g(\alpha^*) \leq g_1(\alpha^*) \leq g_2(\alpha^*) \leq \frac{g'(0)}{2} \alpha^*. \tag{3.39}$$

Now we are in the position to state the main result in this section.

Theorem 3.3.4 *Let the function $g(\alpha)$ be defined by (3.24) with $\Psi(v) \geq \nu_1^{-1}$. Then the step size $\alpha = \alpha^*$ given by (3.30) or the step size $\alpha = \nu_5\sigma^{-(q+1)/q}$ is strictly feasible. Moreover,*

$$g(\alpha) \leq \frac{1}{2}g'(0)\alpha \leq -\frac{\nu_5\nu_1^{(q-1)/(2q)}}{4}\Psi(v)^{(q-1)/(2q)}.$$

In the special case where $\psi(t) = \Upsilon_{p,q}(t)$, given by (2.5), this bound simplifies to

$$g(\alpha) \leq -\min\left\{\frac{1}{12p + 4}, \frac{1}{24q + 16}\right\}\Psi(v)^{(q-1)/(2q)}.$$

Proof It suffices to prove the first result of the theorem. First we observe that for the given step size in the theorem, one has $\alpha \leq \alpha^*$ and thus the strict feasibility of the step size follows directly from the definition of α^*. Now, from (3.39) and (3.32) and by making use of Lemma 1.3.3, we conclude that

$$g(\alpha) \leq -\frac{\nu_5}{4}\sigma^{(q-1)/q} \leq -\frac{\nu_5\nu_1^{(q-1)/(2q)}}{4}\Psi(v)^{(q-1)/(2q)},$$

where the second inequality is implied by (3.10) in Proposition 3.1.5. □

We would like to mention that Theorem 3.3.4 only provides a lower bound for the decreasing value of the proximity. In practical implementation of IPMs, one can always refer to various line-search techniques to find a step size reducing the proximity much more than what we have presented in Theorem 3.3.4.

3.4 COMPLEXITY OF THE ALGORITHM

We come to one of the main targets of this work in this section: to present a complexity result for large-update LO algorithms that is better than the known $\mathcal{O}(nL)$ iteration bound. For this we first derive an upper bound for the number of iterations of the algorithm under the assumption that at each step the damping factor α^* is defined by (3.30) or the damped step size $\alpha = \nu_5 \sigma^{-(q+1)/q}$, as it is used as in Theorem 3.3.4. Note that when either of these two step sizes is applied, then after one Newton step the proximity decreases by at least $(\nu_5/4)(\nu_1 \Psi(v))^{(q-1)/(2q)}$ depending on the current value of the proximity. On the other hand, when the present iterate v enters the neighborhood defined in the algorithm, that is, $\Psi(v) \leq \tau$, then we need to update the parameter μ by $\mu := (1 - \theta)\mu$ and consequently the new point in the v-space becomes $\hat{v} := v/\sqrt{1 - \theta}$. In light of this change, the proximity after one update of the parameter μ might increase. Our next lemma provides an upper bound for the increase of $\Psi(\hat{v})$ after one update of μ.

Lemma 3.4.1 *Suppose that* $\Psi(v) \leq \tau$ *and* $\hat{v} = v/\sqrt{1 - \theta}$. *Then one has*

$$\Psi(\hat{v}) \leq \frac{\nu_2 \tau}{\nu_1 (1 - \theta)^{(p+1)/2}} + \nu_2 \Upsilon'_{p,q}\left((1 - \theta)^{-1/2}\right) \sqrt{\frac{2n\tau}{\nu_1 (1 - \theta)}}$$

$$+ n\nu_2 \Upsilon_{p,q}\left((1 - \theta)^{-1/2}\right).$$

Proof Replacing the parameter ϑ by $1/\sqrt{1 - \theta}$ in inequality (3.14) of Proposition 3.1.5, one obtains the desired inequality.   □

For ease of reference, let us denote by $\psi_0(\theta, \tau, n)$ the right-hand side of the inequality in Lemma 3.4.1. For simplicity, we also denote by $\theta_1 = 1 - \theta$, then one has

$$\psi_0(\theta, \tau, n) = \frac{\nu_2 \tau}{\nu_1 \theta_1^{(p+1)/2}} + \nu_2 \Upsilon'_{p,q}\left(\theta_1^{-1/2}\right) \sqrt{\frac{2n\tau}{\nu_1 \theta_1}} + n\nu_2 \Upsilon_{p,q}\left(\theta_1^{-1/2}\right). \quad (3.40)$$

It is easy to check that if $\tau = \mathcal{O}(n)$ and $\theta \in (0, 1)$ is a fixed constant independent of n, then one has $\psi_0(\theta, \tau, n) = \mathcal{O}(n)$. Note that the expression on the right-hand side of (3.40) is exponential with respect to the parameter p. Therefore, whenever $\tau = \mathcal{O}(n)$, to keep the proximity $\psi_0(\theta, \tau, n) = \mathcal{O}(n)$ for fixed $\theta \in (0, 1)$, one should choose a reasonably small p.

Combining Lemma 3.4.1, Proposition 1.3.2 and Theorem 3.3.4, we obtain the following result as an upper bound for the number of inner iterations in the algorithm.

Lemma 3.4.2 *Suppose $\Phi(x, s, \mu) \leq \tau$ and $\tau \geq \nu_1^{-1}$. Then after an update of the barrier parameter, no more than*

$$\left\lceil \frac{8q\nu_1^{-(q-1)/(2q)}}{\nu_5(q+1)}(\psi_0(\theta, \tau, n))^{(q+1)/(2q)} \right\rceil$$

inner iterations are needed to recenter. In the special case where $\psi(t) = \Upsilon_{p,q}(t)$, given by (2.5), the above bound simplifies to

$$\left\lceil \frac{8q \max\{3p+1, 6q+4\}}{q+1}(\psi_0(\theta, \tau, n))^{(q+1)/(2q)} \right\rceil.$$

We are now ready to bound the total number of iterations.

Theorem 3.4.3 *If $\tau \geq \nu_1^{-1}$, the total number of iterations required by the primal-dual Newton algorithm is not more than*

$$\left\lceil \frac{8q\nu_1^{-(q-1)/2q}}{\nu_5(q+1)}(\psi_0(\theta, \tau, n))^{(1+q)/2q} \right\rceil \left\lceil \frac{1}{\theta} \log \frac{n}{\varepsilon} \right\rceil.$$

In the special case when $\psi(t) = \Upsilon_{p,q}(t)$, given by (2.5), this bound (with $\nu_1 = \nu_2 = 1$) simplifies to

$$\left\lceil \frac{8q \max\{3p+1, 6q+4\}}{q+1}(\psi_0(\theta, \tau, n))^{(q+1)/2q} \right\rceil \left\lceil \frac{1}{\theta} \log \frac{n}{\varepsilon} \right\rceil.$$

Proof The number of barrier parameter updates is given by (cf. Lemma II.17, page 116, in [98])

$$\left\lceil \frac{1}{\theta} \log \frac{n}{\varepsilon} \right\rceil.$$

Multiplication of this number by the bound for the number of inner iterations in Lemma 3.4.2 yields the theorem. $\qquad\square$

For large-update IPMs, omitting the round off brackets in Theorem 3.4.3 does not change the order of magnitude of the iteration bound. From the definition of $\psi_0(\theta, \tau, n)$, (3.40), one can easily show that there exists a constant $\nu_6 > 0$ depending on the function $\psi(t)$ and θ such that $\psi_0(\theta, \tau, n) \leq n\nu_6$. Hence we may safely consider the following expression as an upper bound for the number of iterations:

$$\mathcal{O}\left(n^{(q+1)/(2q)} \log \frac{n}{\varepsilon}\right).$$

Let us consider a specific case in which the function $\psi(t)$ is defined by

$$\psi(t) = \frac{1}{2}(t^2 - 1) + \frac{1}{q-1}(t^{1-q} - 1).$$

Then the constants are $\nu_1 = 1$, $\nu_2 = q$ and the search direction in this mono-graph reduces to the same search direction introduced in [90], while the complexity presented here is better than by the same authors in [90] and the limitation $q \in [1, 3]$ is also removed.

Let us elaborate a little more on another interesting class of *self-regular* functions: $\Upsilon_{p,q}(t)$ defined by (2.5) with $p, q \geq 1$. Since the function $\Upsilon_{p,q}(t)$ satisfies SR1 and SR2 with $\nu_1 = \nu_2 = 1$ for any $p, q \geq 1$, we can specify the parameters p and q. In particular, we can choose $p = 2$, $q = \log n$. Then the last statement of Theorem 3.4.3 provides us with the following upper bound for the total number of iterations:

$$\mathcal{O}\left(\sqrt{n} \log n \log \frac{n}{\varepsilon}\right).$$

This gives to date the best complexity bound known for large-update methods. Note that the order of magnitude does not change if we take $\tau = n$, or $\tau = \mathcal{O}(n)$, which is in practice more attractive and quite close to what is implemented in most of the existing IPM solvers [7]. We would like to remind the reader that theoretically we can also choose p very big. However, when we take $\tau = \mathcal{O}(n)$, which means the algorithm indeed works in a large neighbor-hood, then one should use only a small parameter p (for instance, $p = \mathcal{O}(1)$) because otherwise after a large update of μ, the proximity $\Psi(v)$ might become much larger than $\mathcal{O}(n)$.

For the small-update method, if we choose θ sufficiently small such that

$$\frac{1}{\sqrt{1-\theta}} \leq 1 + \frac{\nu_3}{\sqrt{n}},$$

then by (3.16) we know that there exists a constant ν_7 such that

$$\psi_0(\theta, \tau, n) \leq \nu_7.$$

Theorem 3.4.3 then shows that the small-update method still has an $\mathcal{O}(\sqrt{n} \log(n/\varepsilon))$ iteration bound.

3.5 RELAXING THE REQUIREMENT ON THE PROXIMITY FUNCTION

At the end of this chapter, we consider the possibility of weakening the requirements on the function $\psi(t)$. More concretely, the question to be addressed is whether we can establish similar complexity results for some new algorithms in which the search direction is defined by a certain proximity that only satisfies condition SR1. By using several properties of the kernel functions, we present an affirmative answer to this question.

Let us first review how we have obtained the estimate of the proximity after one Newton step. Let $g(\alpha), g_1(\alpha)$ be the functions defined by (3.24) and (3.25) respectively. The role of condition SR2 is crucial in establishing relation (3.25), which enables us to bound the function $g(\alpha)$ from above by $g_1(\alpha)$. A key step in evaluating $g_1(\alpha)$ is to prove inequality (3.26) for the second derivative $g_1''(\alpha)$ in Lemma 3.3.2. Since $g'(0) = g_1'(0)$, it is not difficult to see that if we can derive directly some inequality for the second derivative $g''(\alpha)$ analogous to (3.26), then in a similar vein we can estimate the proximity after one step (or equivalently $g(\alpha)$) without requiring relation (3.25). In such a situation, condition SR2 becomes superfluous. In what follows we give such an estimate for $g''(\alpha)$.

Lemma 3.5.1 *Let the function $g(\alpha)$ be defined by (3.24) and suppose $\psi(t) \in \Omega_1$. Then for any $\alpha \in [0, \overline{\alpha}/2)$, we have*

$$g''(\alpha) \le \frac{3\nu_2\sigma^2}{2}\left(1 + \frac{\nu_2}{\nu_1}\right)((v_{\max} + \alpha\sigma)^{p-1} + (v_{\min} - \alpha\sigma)^{-q-1}). \quad (3.41)$$

Moreover, if $\psi(t) \in \Omega_1$ with parameter $p \ge 3$, then for any $\alpha \in [0, \overline{\alpha})$,

$$g''(\alpha) \le \frac{\nu_2\sigma^2}{2}\left(1 + \frac{\nu_2}{\nu_1}\right)((v_{\max} + \alpha\sigma)^{p-1} + (v_{\min} - \alpha\sigma)^{-q-1}). \quad (3.42)$$

Proof First recall that $v_+ = (v + \alpha d_x)^{1/2}(v + \alpha d_s)^{1/2}$. From (3.24) we obtain

$$g''(\alpha) = \sum_{i=1}^{n} \frac{d^2\psi(v_{+i})}{d\alpha^2}$$

$$= \frac{1}{4}\sum_{i=1}^{n}\left(\frac{v_i + \alpha[d_s]_i}{v_i + \alpha[d_x]_i}[d_x]_i^2 + 2[d_x]_i[d_s]_i + \frac{v_i + \alpha[d_x]_i}{v_i + \alpha[d_s]_i}[d_s]_i^2\right)\psi''(v_{+i})$$

$$- \frac{1}{4}\sum_{i=1}^{n}\left(\frac{v_i + \alpha[d_s]_i}{v_i + \alpha[d_x]_i}[d_x]_i^2 - 2[d_x]_i[d_s]_i + \frac{v_i + \alpha[d_x]_i}{v_i + \alpha[d_s]_i}[d_s]_i^2\right)\left(\frac{\psi(v_{+i})}{v_{+i}}\right).$$

It is trivial to see that

$$[d_x]_i[d_s]_i \le \frac{1}{2}\left(\frac{v_i + \alpha[d_s]_i}{v_i + \alpha[d_x]_i}[d_x]_i^2 + \frac{v_i + \alpha[d_x]_i}{v_i + \alpha[d_s]_i}[d_s]_i^2\right) \quad \forall i = 1, ..., n.$$

Hence, by using condition SR1 and Lemma 2.2.1, one has

$$g''(\alpha) \leq \frac{\nu_2}{2}\left(1+\frac{\nu_2}{\nu_1}\right)\sum_{i=1}^{n}\left(\frac{v_i+\alpha[d_s]_i}{v_i+\alpha[d_x]_i}[d_x]_i^2 + \frac{v_i+\alpha[d_x]_i}{v_i+\alpha[d_s]_i}[d_s]_i^2\right)\left(v_{+i}^{p-1}+v_{+i}^{-q-1}\right)$$

$$\leq \frac{\nu_2}{2}\left(1+\frac{\nu_2}{\nu_1}\right)\left((v_{\max}+\alpha\sigma)^{p-1}+(v_{\min}-\alpha\sigma)^{-q-1}\right)$$

$$\times \sum_{i=1}^{n}\left(\frac{1+\alpha[\bar{d}_s]_i}{1+\alpha[\bar{d}_x]_i}[d_x]_i^2 + \frac{1+\alpha[\bar{d}_x]_i}{1+\alpha[\bar{d}_s]_i}[d_s]_i^2\right), \tag{3.43}$$

where the last inequality is true, because from (3.27) and (3.28) one can easily derive the relation

$$v_{\min}-\alpha\sigma \leq v_{+i} \leq v_{\max}+\alpha\sigma, \quad i \in I. \tag{3.44}$$

If $\alpha \in [0,\bar{\alpha}/2]$, it is easy to see that for any $i \in I$,

$$0 \leq \frac{1+\alpha[\bar{d}_x]_i}{1+\alpha[\bar{d}_s]_i}, \frac{1+\alpha[\bar{d}_s]_i}{1+\alpha[\bar{d}_x]_i} \leq 3.$$

This relation, together with (3.43), (3.9) and (3.20), gives the first statement of the lemma.

To prove the second conclusion of the lemma, we observe that from the first inequality in (3.43) one can derive

$$g''(\alpha) \leq \frac{\nu_2}{2}\sum_{i=1}^{n}\left((v_i+\alpha[d_s]_i)^2[v_+]_i^{p-3} + [v_+]_i^{-q+1}(v_i+\alpha[d_x]_i)^{-2}\right)[d_x]_i^2$$

$$+\frac{\nu_2^2}{2\nu_1}\sum_{i=1}^{n}\left([v_+]_i^{p-3}(v_i+\alpha[d_x]_i)^2 + [v_+]_i^{-q+1}(v_i+\alpha[d_s]_i)^{-2}\right)[d_s]_i^2$$

$$\leq \frac{\nu_2}{2}\left(1+\frac{\nu_2}{\nu_1}\right)\left((v_{\max}+\alpha\sigma)^{p-1}+(v_{\min}-\alpha\sigma)^{-q-1}\right)\sum_{i=1}^{n}\left([d_x]_i^2 + [d_s]_i^2\right)$$

$$=\frac{\nu_2\sigma^2}{2}\left(1+\frac{\nu_2}{\nu_1}\right)\left((v_{\max}+\alpha\sigma)^{p-1}+(v_{\min}-\alpha\sigma)^{-q-1}\right),$$

where the second inequality can be obtained by using (3.27), (3.28) and (3.44), and the last equality follows from the definition of σ (3.9) and (3.20). □

The above lemma indicates that at least for linear programming, condition SR2 can be waived. We can use any function belonging to Ω_1 to define the proximity and subsequently the search direction. By applying Lemma 2.2.1

and Lemma 3.5.1 and following arguments akin to what we have done in Sections 3.3 and 3.4, for large-update IPMs one can still get the following expression as an upper bound on the number of iterations:

$$\mathcal{O}\left(n^{(q+1)/(2q)}\log\frac{n}{\varepsilon}\right).$$

The details are left to the interested reader.

Chapter 4

Interior-Point Methods for Complementarity Problems Based on Self-Regular Proximities

This chapter extends the approach for solving LO problems in the previous chapter to the case of linear and nonlinear $P_*(\kappa)$ complementarity problems (LCPs and NCPs). First, several elementary results about $P_*(\kappa)$ mappings are provided. To establish global convergence of the algorithm, a new smoothness condition for the mapping is introduced. This condition is closely related to the relative Lipschitz condition and applicable to nonmonotone mappings. New search directions for solving the underlying problem are proposed based on self-regular proximities. We show that if a strictly feasible starting point is available and the mapping involved satisfies a certain smoothness condition, then these new IPMs for solving $P_*(\kappa)$ CPs have polynomial iteration bounds similar to those of their LO cousins.

4.1 INTRODUCTION TO CPS AND THE CENTRAL PATH

The classical complementarity problem (CP) is to find an x in $\Re^n$ such that

$$x \geq 0, \quad f(x) \geq 0, \quad x^{\mathrm{T}}f(x) = 0,$$

where $f(x) = (f_1(x), \ldots, f_n(x))^{\mathrm{T}}$ is a mapping from $\Re^n$ into itself. To be more specific, we also call it a linear complementarity problem (LCP) if the involved mapping $f(x)$ is affine, that is, $f(x) = Mx + c$ for some $M \in \Re^{n \times n}$, $c \in \Re^n$; otherwise we call it a nonlinear complementarity problem (NCP) when $f(x)$ is nonlinear.

In view of the nonnegativity requirements on x and $f(x)$ at the solution point of an NCP, since $x^{\mathrm{T}}f(x) = \sum_{i=1}^{n} x_i f_i(x)$, a standard CP can be equivalently stated as

$$x \geq 0, \quad f(x) \geq 0, \quad xf(x) = 0,$$

where $xf(x)$ denotes the vector whose ith component is $x_i f_i(x)$. From a purely mathematical point of view, finding a solution of a CP is also equivalent to identifying a global minimizer of the following optimization problem:

$$\min x^{\mathrm{T}}f(x)$$

$$\text{s.t.} \qquad x \geq 0, \quad f(x) \geq 0,$$

or locating a solution of the nonsmooth equation system (or equilibrium problem) defined as follows:

$$\min\{x, f(x)\} = 0.$$

The above diverse but equivalent formulations of a CP illustrate the broad range of associations of CPs with different areas. Indeed, at its infancy in the 1960s, the LCP was closely associated with LO and quadratic optimization problems. Now it is known that CPs cover fairly general classes of mathematical programming problems with many applications in engineering, economics, and science. For instance, by exploiting the first-order optimality conditions of the underlying optimization problem, we can model any general convex optimization problem satisfying certain constraint qualifications (e.g., the Slater constraint qualification [67]) as a monotone CP. Closely related to CPs is a large class of problems: variational inequality problems (VIPs). VIPs are widely used in the study of equilibrium in, for example, economics, transportation planning and game theory. As a result of its wide association with optimization and equilibrium problems, the study of CPs has attracted much attention from researchers in different fields such as operations research, mathematics, computer science, economics and engineering ever since its introduction. Several monographs [15,44] and surveys [22,37,87] have documented the basic theory, algorithms and applications of CPs and their role in optimization theory. Many classical numerical algorithms for solving CPs are based on approaches for optimization problems or equation systems.

Besides their many meaningful applications, CPs contain one common feature that is crucial to the study of general mathematical programming and equilibrium problems. This is the concept of complementarity. As we have seen already in the previous chapters for LO, the concept of complementarity plays an important role in understanding the nature of the underlying problem as well as in the design and analysis of numerical algorithms for solving the problem.

In the present chapter, among the many facets of the research on CPs we focus on efficient IPMs for solving CPs. As one might expect, most IPMs for CPs follow a similar recipe to that for LO: using variants of Newton-type methods to trace the central path of the underlying problem and approximate the solution set of the problem until the parameter defining the central path decreases to zero. A close look at the IPM literature tells us that the first IPM for LCPs was due to Kojima, Mizuno and Yoshise [60] and their algorithm originated from the primal-dual IPMs for LO. Later Kojima, Megiddo, Noma and Yoshise [58] set up a framework of IPMs for tracing the central path of a class of LCPs. Independent of the works by this Japanese group is the paper by Monteiro and Adler [76] who proposed an IPM for convex quadratic optimization problems. Monteiro and Adler's algorithm could indeed be applied to LCPs. Similar results can also be found in [72a, 72b]. Since then, the study of IPMs for CPs has been carried out in a way parallel to that for LO. A general and unified analysis about path-following methods for VIPs and CPs was given by Nesterov and Nemirovskii in [83]. The survey by Yoshise [129] gave a comprehensive review of the major developments in IPMs for CPs and listed many available references up to that time.

Even in the case of general convex optimization, the global convergence of IPMs has been established only for the class of problems that satisfy certain Lipschitz conditions such as the self-concordant condition posed by Nesterov and Nemirovskii [83], the relative Lipschitz condition introduced by Jarre [50,51], and the scaled Lipschitz condition by Zhu [136]. For CPs, Jansen et al. [45,49] introduced a smoothness condition that can be viewed as a straightforward extension of the scaled Lipschitz condition. In this chapter, to establish the complexity of our algorithm, we introduce a new smoothness condition for the underlying CP. This new condition can be regarded as a generalization of Jarre's condition.

To continue our discussion, we first introduce some concepts related to classes of matrices [58]. For simplicity, we assume throughout the present chapter that κ is a nonnegative constant.

Definition 4.1.1 *A matrix $M \in \Re^{n \times n}$ is said to be a $P_*(\kappa)$ matrix if and only if*

$$(1 + 4\kappa) \sum_{i \in I_+(x)} x_i[Mx]_i + \sum_{i \in I_-(x)} x_i[Mx]_i \geq 0 \quad \forall x \in \Re^n,$$

where

$$I_+(x) = \{i \in I : x_i[Mx]_i \geq 0\}, \quad I_-(x) = \{i \in I : x_i[Mx]_i < 0\}.$$

We remind the reader that the index sets $I_+(x)$ and $I_-(x)$ depend not only on $x \in \Re^n$ but also on the matrix M. The class of $P_*(\kappa)$ matrices includes as a specific case the class of positive semidefinite matrices, for which the constant $\kappa = 0$. For more discussion about the relations among the class of $P_*(\kappa)$ matrices and other classes of matrices, we refer to [58].

We denote by P_* the union of all $P_*(\kappa)$ matrices with $\kappa \geq 0$. The notion of a $P_*(\kappa)$ mapping is defined as follows.

Definition 4.1.2 *A mapping $f(x) : \Re^n \to \Re^n$ is said to be a $P_*(\kappa)$ mapping if for any $x \neq y \in \Re^n$, the relation*

$$(1 + 4\kappa) \sum_{i \in I_+^f(x,y)} (x_i - y_i)(f_i(x) - f_i(y)) + \sum_{i \in I_-^f(x,y)} (x_i - y_i)(f_i(x) - f_i(y)) \geq 0$$

(4.1)

holds, where

$$I_+^f(x,y) = \{i \in I : (x_i - y_i)(f_i(x) - f_i(y)) \geq 0\},$$

$$I_-^f(x,y) = \{i \in I : (x_i - y_i)(f_i(x) - f_i(y)) < 0\}.$$

The mapping $f(x)$ is said to be a strict $P_*(\kappa)$ mapping if inequality (4.1) holds strictly for any $x \neq y \in \Re^n$.

It follows directly from the above two definitions that if $f(x) = Mx + c$, then $f(x)$ is a $P_*(\kappa)$ mapping if and only if its Jacobian matrix M is a $P_*(\kappa)$ matrix.

In this chapter we elaborate mainly on IPMs for solving $P_*(\kappa)$ CPs. The class of $P_*(\kappa)$ CPs is a rather general class of CPs that covers CPs with P (see Definition 4.2.2) and monotone mappings. In the case of LCPs, it reduces to the class of LCPs introduced by [58], which is to date the largest set of LCPs that could be solved by IPMs in polynomial time. As we mentioned early at the beginning of this section, many convex optimization problems can be reformulated as monotone CPs: a subclass of $P_*(\kappa)$ CPs.

To be consistent with the notation in the LO chapter, we introduce an artificial vector s and rewrite the standard CP as follows:

$$s = f(x), \quad xs = 0, \quad x \geq 0, \quad s \geq 0.$$

(4.2)

The central path for the above CP is defined as the solution of the following parameterized system of nonlinear equations:

$$s = f(x), \quad xs = \mu e, \tag{4.3}$$

where $\mu > 0$ is a positive parameter and e is the vector of ones. Even before the publication of Karmarkar's pioneering work [52], McLinden had built some fundamental results for a certain class of CPs in his remarkable paper [68] and his results have proved to be very helpful for the study of IPMs for CPs. In [58] Kojima et al. showed that if a $P_*(\kappa)$ LCP is strictly feasible, then the central path of the underlying LCP is well defined. In the case of LCPs, the results presented in [58] extended those in [68]. Kojima et al. [57] discussed the existence of the central path of a monotone CP under the assumption that the problem is strictly feasible. The analysis in [57, Section 3] can be easily extended to the case where $f(x)$ is a $P_*(\kappa)$ mapping (see [93]).

Throughout this chapter we further assume that the CP under consideration is strictly feasible. This is a general requirement in the literature of feasible IPMs for CPs. For monotone CPs (see Section 7.2), by using an augmented homogeneous model described by Andersen and Ye in [9], we can always get a strictly feasible point for the reconstructed CP. For $P_*(\kappa)$ LCPs, one can apply the big-$\mathcal{M}$ method introduced in the monograph [58] to get a strictly feasible initial point. However, as observed by Peng, Roos and Terlaky [88], Andersen and Ye's homogeneous model cannot be applied to a $P_*(\kappa)$ CP because there is no guarantee that the reformulated CP is still in the class of $P_*(\kappa)$ CPs.

Recently Peng, Terlaky and Yoshise proposed a new augmented model that can transfer any CP into a stricly feasible CP. The properties of the central path of the embedded problem and the relation between the solution sets of the original and the embedded CP are also explored.

By following a procedure analogous to the discussion in [57, Section 3] (see also the discussion in [93, Section 2]), we can readily prove the following result.

Theorem 4.1.3 *Suppose that problem (4.2) has a strictly feasible solution and that $f(x)$ is a $P_*(\kappa)$ mapping. Then for any positive μ, system (4.3) has a unique solution.*

We mention that in [134,135], the authors also discussed the existence of the central path and solvability of $P_*(\kappa)$ CPs and showed that if $f(x)$ is a $P_*(\kappa)$ mapping, then the strict feasibility of the underlying CP implies that the central path exists and the solution set of the problem is nonempty and bounded.

IPMs track the homotopy trajectory defined by (4.3) appropriately and approximate the solution set of the underlying CP as μ goes to zero.

4.2 PRELIMINARY RESULTS ON $P_*(\kappa)$ MAPPINGS

As we observed in the introduction of this chapter, an affine mapping $f(x) = Mx + c$ is a $P_*(\kappa)$ mapping if and only if its Jacobian matrix M is a $P_*(\kappa)$ matrix. In other words, an affine $P_*(\kappa)$ mapping can be defined according to Definition 4.1.2 or by requiring that its Jacobian matrix M is a $P_*(\kappa)$ matrix. Possibly because of this fact, there exist two different ways for defining a $P_*(\kappa)$ mapping in the CP literature and this unfortunate difference came from diverse researchers who were studying various aspects of $P_*(\kappa)$ CPs. For instance, in some references (see [134,135]) dealing with the existence of the central path and the feasibility of the considered CP, a $P_*(\kappa)$ mapping is defined based on Definition 4.1.2, because such a definition can be applied directly to the study of the topics of the references. However, in some other references such as [49] dealing with IPMs for CPs, because we always need to solve a linear system involving the Jacobian matrix of the mapping, it is required that the Jacobian matrix be a $P_*(\kappa)$ matrix to ensure that the system is well defined.

The issue we want to address in the sequel is whether these two different definitions are identical under certain condition and thus close the above-mentioned gap in the definition of $P_*(\kappa)$ mappings. As we see later, this is far from a trivial task.[1] We start with some basic definitions of classes of matrices [15].

Definition 4.2.1 *A matrix $M \in \Re^{n \times n}$ is said to be a P (or P_0) matrix if and only if for any $x \neq 0 \in \Re^n$, there exists at least one index $i \in I$ such that $x_i(Mx)_i > 0$ (or $x_i(Mx)_i \geq 0$).*

From the above definition and (4.1.1), one can easily see that a $P_*(\kappa)$ matrix is obviously a P_0 matrix. On the other hand, we can prove that any P matrix is a $P_*(\kappa)$ matrix. Since P_* is the union of all $P_*(\kappa)$ matrices with $\kappa \geq 0$, we therefore get the relation $P \subset P_* \subset P_0$. More details about various relations among these classes of matrices can be found in [15,58].

The following technical result about P and P_0 matrices is used in our later discussion.

Lemma 4.2.2 *If M is a P matrix, then there exists a vector x such that*

$$Mx > 0, \quad x > 0.$$

If M is a P_0 matrix, then there exists a nonzero vector x such that

$$Mx \geq 0, \quad x \geq 0.$$

---

[1] In [79], Moré and Rheinboldt gave an example showing that the Jacobian matrix of a continuously differentiable P mapping (see Definition 4.2.3) might not be a P matrix. After the release of [93], Professor Potra called the attention of the authors to [62] for similar discussion.

Proof The first statement of the lemma is precisely the same as Corollary 3.3.5 in [15], so its proof is omitted here. To prove the second statement of the lemma, we observe that if M is a P_0 matrix, then for any $\varepsilon > 0$ the matrix $M + \varepsilon E$ is a P matrix, where we denote by E the identity matrix in $\Re^{n \times n}$. Thus, from the first statement of the lemma, we know that for any $\varepsilon > 0$, there exists a vector $x_\varepsilon > 0$ with $\|x_\varepsilon\| = 1$ such that

$$(M + \varepsilon E)x_\varepsilon > 0, \quad x_\varepsilon > 0.$$

Therefore, there must exist an accumulation point x^* of the sequence x_{ε_k} as ε_k decreases to zero. By taking limits if necessary, one can see that

$$Mx^* \geq 0, \quad x^* \geq 0,$$

which completes the proof of the second result of the lemma. $\qquad\square$

We proceed by introducing a specific subclass of $P_*(\kappa)$ mappings and matrices.

Definition 4.2.3 *Let β be a nonnegative number. A mapping $f(x) : \Re^n \to \Re^n$ is said to be a $P_*(\kappa, \beta)$ mapping if for any $x \neq y \in \Re^n$,*

$$(1 + 4\kappa) \sum_{i \in I_+^f(x,y)} (x_i - y_i)(f_i(x) - f_i(y)) + \sum_{i \in I_-^f(x,y)} (x_i - y_i)(f_i(x) - f_i(y))$$

$$\geq \beta \|x - y\|^2,$$

where

$$I_+^f(x, y) = \{i \in I : (x_i - y_i)(f_i(x) - f_i(y)) \geq 0\},$$

$$I_-^f(x, y) = \{i \in I : (x_i - y_i)(f_i(x) - f_i(y)) < 0\}.$$

An immediate consequence of this definition is that a $P_*(\kappa, \beta)$ mapping with $\beta > 0$ is a strict $P_*(\kappa)$ mapping. We mention that when $\beta = 0$, the class of $P_*(\kappa, 0)$ mappings coincides with the class of $P_*(\kappa)$ mappings.

Definition 4.2.4 *Let β be a nonnegative number. A matrix M is said to be a $P_*(\kappa, \beta)$ matrix if for any $x \in \Re^n$,*

$$(1 + 4\kappa) \sum_{i \in I_+(x)} x_i(Mx)_i + \sum_{i \in I_-(x)} x_i(Mx)_i \geq \beta \|x\|^2,$$

where

$$I_+(x) = \{i \in I : x_i(Mx)_i \geq 0\}$$

$$I_-(x) = \{i \in I : x_i(Mx)_i < 0\}.$$

Our next result characterizes the interrelation between $P_*(\kappa)$ and $P_*(\kappa,\beta)$ mappings.

Lemma 4.2.5 *A mapping $f(x) : \Re^n \to \Re^n$ is a $P_*(\kappa)$ mapping if and only if for any positive $\beta > 0$, the mapping $f_\beta(x) = f(x) + \beta x$ is a $P_*(\kappa,\beta)$ mapping.*

Proof The necessary part of the lemma is trivial. First we observe that if $f(x)$ is a $P_*(\kappa)$ mapping with $\kappa \geq 0$, then for any $x,y \in \Re^n$ we know that the set $I^f_+(x,y)$ is nonempty. Further, it is easy to see that for any $\beta > 0$ the inclusions $I^f_+(x,y) \subseteq I^{f_\beta}_+(x,y)$ and $I^{f_\beta}_-(x,y) \subseteq I^f_-(x,y)$ hold. For simplification of expression, in the remaining part of the proof of this lemma we also omit the variable pair (x,y) in the index sets. Therefore it follows directly that

$$(1 + 4\kappa) \sum_{i \in I^{f_\beta}_+} (x_i - y_i)(f^i_\beta(x) - f^i_\beta(y)) + \sum_{i \in I^{f_\beta}_-} (x_i - y_i)(f^i_\beta(x) - f^i_\beta(y))$$

$$\geq (1 + 4\kappa) \sum_{i \in I^f_+} (x_i - y_i)(f^i_\beta(x) - f^i_\beta(y)) + \sum_{i \in I^f_-} (x_i - y_i)(f^i_\beta(x) - f^i_\beta(y))$$

$$\geq \beta\|x - y\|^2 + (1 + 4\kappa) \sum_{i \in I^f_+} (x_i - y_i)(f_i(x) - f_i(y))$$

$$+ \sum_{i \in I^f_-} (x_i - y_i)(f_i(x) - f_i(y))$$

$$\geq \beta\|x - y\|^2,$$

where the first two inequalities follow from the assumption that $\kappa \geq 0$ and the fact that $I^{f_\beta}_+(x,y)$ is nonempty, and the last inequality is given by the definition of a $P_*(\kappa)$ mapping.

To prove the sufficient part of the lemma, let us assume that $f_\beta(x)$ is a $P_*(\kappa,\beta)$ mapping for any sufficiently small $\beta > 0$. Suppose the converse that the statement of the lemma is false, that is, $f(x)$ is not a $P_*(\kappa)$ mapping. Then, from Definition 4.1.2, we deduce that there exist $x,y \in \Re^n$ such that

$$(1 + 4\kappa) \sum_{i \in I^f_+} (x_i - y_i)(f_i(x) - f_i(y)) + \sum_{i \in I^f_-} (x_i - y_i)(f_i(x) - f_i(y)) < 0.$$

Let us denote

$$\beta_0 = \frac{(1 + 4\kappa) \sum_{i \in I^f_+} (x_i - y_i)(f_i(x) - f_i(y)) + \sum_{i \in I^f_-} (x_i - y_i)(f_i(x) - f_i(y))}{\|x - y\|^2},$$

$$\beta_1 = \max_{i \in I_-^f} \left\{ \frac{(x_i - y_i)(f_i(x) - f_i(y))}{\|x - y\|^2} \right\}.$$

One can easily verify that both β_0 and β_1 are negative. Let

$$\beta_2 = \frac{1}{2} \min \left\{ \frac{-\beta_0}{1 + 4\kappa}, -\beta_1 \right\}.$$

Obviously $\beta_2 > 0$. Let us define $f_{\beta_2}(x) = f(x) + \beta_2 x$ for any $x \in \Re^n$. For this specific mapping f_{β_2}, it is straightforward to check that $I_-^{f_{\beta_2}}(x, y) = I_-^f(x, y)$ and hence $I_+^{f_{\beta_2}}(x, y) = I_+^f(x, y)$. Therefore,

$$(1 + 4\kappa) \sum_{i \in I_+^{f_{\beta_2}}} (x_i - y_i)(f_{\beta_2}^i(x) - f_{\beta_2}^i(y)) + \sum_{i \in I_-^{f_{\beta_2}}} (x_i - y_i)(f_{\beta_2}^i(x) - f_{\beta_2}^i(y))$$

$$= (1 + 4\kappa) \sum_{i \in I_+^f} (x_i - y_i)(f_{\beta_2}^i(x) - f_{\beta_2}^i(y)) + \sum_{i \in I_-^f} (x_i - y_i)(f_{\beta_2}^i(x) - f_{\beta_2}^i(y))$$

$$= (1 + 4\kappa) \sum_{i \in I_+^f} (x_i - y_i)(f_i(x) - f_i(y)) + \sum_{i \in I_-^f} (x_i - y_i)(f_i(x) - f_i(y))$$

$$+ (1 + 4\kappa)\beta_2 \sum_{i \in I_+^f} (x_i - y_i)^2 + \beta_2 \sum_{i \in I_-^f} (x_i - y_i)^2$$

$$\leq \beta_0 \|x - y\|^2 + (1 + 4\kappa)\beta_2 \|x - y\|^2 \leq \frac{\beta_0}{2} \|x - y\|^2 < 0,$$

where the first inequality is true because $\kappa \geq 0$, and the last inequality follows from the choice of β_2. The above discussion means that the mapping $f_{\beta_2}(x)$ is not a $P_*(\kappa)$ mapping and thus also not a $P_*(\kappa, \beta)$ mapping. This contradicts our assumption that $f_\beta(x)$ is a $P_*(\kappa, \beta)$ mapping for any positive $\beta > 0$. Hence $f(x)$ must be a $P_*(\kappa)$ mapping, which completes the proof of the lemma. $\square$

Recall that when $\beta \geq 0$, any $P_*(\kappa, \beta)$ mapping is naturally a $P_*(\kappa)$ mapping. We thus obtain the following result as a refinement of Lemma 4.2.5.

Corollary 4.2.6 *A mapping $f(x) : \Re^n \to \Re^n$ is a $P_*(\kappa)$ mapping if and only if for any positive $\beta > 0$, the mapping $f_\beta(x) = f(x) + \beta x$ is a $P_*(\kappa)$ mapping.*

One can prove the following results for $P_*(\kappa)$ and $P_*(\kappa, \beta)$ matrices similarly, by specifying the mapping $f(x)$ to $f(x) = Mx + c$.

Corollary 4.2.7 *A matrix $M \in \Re^{n \times n}$ is a $P_*(\kappa)$ matrix if and only if for any positive $\beta > 0$, the matrix $M + \beta E$ is a $P_*(\kappa)$ (or $P_*(\kappa, \beta)$) matrix.*

We now present some relations between a differentiable $P_*(\kappa)$ mapping and its Jacobian matrix $\nabla f(x)$.

Lemma 4.2.8 *Suppose that $f(x) : \Re^n \to \Re^n$ is continuously differentiable and $\beta > 0$. If $f(x)$ is a $P_*(\kappa)$ (or $P_*(\kappa, \beta)$) mapping, then for any $x \in \Re^n$, $\nabla f(x)$ is a $P_*(\kappa)$ (or $P_*(\kappa, \beta)$) matrix.*

Proof We consider first the case for $P_*(\kappa, \beta)$ mappings. To prove the statement of the lemma, for any $x, u \in \Re^n$, let us consider a sequence

$$\left\{ x_u^j := x + \frac{1}{j} u : \; j = 1, 2, \dots \right\}.$$

Since $f(x)$ is a $P_*(\kappa, \beta)$ mapping, there exist two sequences of index sets $I_+^j := I_+^f(x_u^j, x)$ and $I_-^j := I_-^f(x_u^j, x)$ such that

$$(1 + 4\kappa) \sum_{i \in I_+^j} \frac{1}{j} u_i \big(f_i(x_u^j) - f_i(x) \big) + \sum_{i \in I_-^j} \frac{1}{j} u_i \big(f_i(x_u^j) - f_i(x) \big) \geq \frac{\beta}{j^2} \|u\|^2.$$

By the finiteness of I, there exist two index sets $I_+'(x, u)$ and $I_-'(x, u)$ and a subsequence J such that for all $j \in J$, $I_+^j = I_+'(x, u)$ and $I_-^j = I_-'(x, u)$ hold. Therefore, for any $j \in J$, we have

$$\frac{1}{j} u_i \big(f_i(x_u^j) - f_i(x) \big) \geq 0 \quad \forall i \in I_+'(x, u),$$

$$\frac{1}{j} u_i \big(f_i(x_u^j) - f_i(x) \big) < 0 \quad \forall i \in I_-'(x, u),$$

and

$$(1 + 4\kappa) \sum_{i \in I_+'(x,u)} \frac{1}{j} u_i \big(f_i(x_u^j) - f_i(x) \big) + \sum_{i \in I_-'(x,u)} \frac{1}{j} u_i \big(f_i(x_u^j) - f_i(x) \big) \geq \frac{\beta}{j^2} \|u\|^2.$$

Taking the limit $j \to \infty$ for $j \in J$, we obtain

$$u_i [\nabla f(x) u]_i \geq 0 \; (i \in I_+'(x, u)), \quad u_i [\nabla f(x) u]_i \leq 0 \; (i \in I_-'(x, u))$$

and

$$(1 + 4\kappa) \sum_{i \in I_+'(x,u)} u_i [\nabla f(x) u]_i + \sum_{i \in I_-'(x,u)} u_i [\nabla f(x) u]_i \geq \beta \|u\|^2,$$

which implies that $\nabla f(x)$ is a $P_*(\kappa, \beta)$ matrix. The proof for $P_*(\kappa)$ mappings follows similarly.   □

In what follows we consider a converse case of the above lemma; we discuss the properties of a continuously differentiable mapping $f(x)$ under the condition that $\nabla f(x)$ is a $P_*(\kappa, \beta)$ matrix for any $x \in \Re^n$.

Lemma 4.2.9 *Suppose that $f(x) : \Re^n \rightarrow \Re^n$ is continuously differentiable. Suppose that for any $x \in \Re^n$, the Jacobian matrix $\nabla f(x)$ is a $P_*(\kappa, \beta)$ matrix with $\beta > 0$. Then $f(x)$ is a strict $P_*(\kappa)$ mapping.*

Proof The proof follows a similar recipe to that in [79] for P mapping. For self-completeness, we give a detailed proof here. The proof is inductive. We first observe that the result is trivial if $n = 1$. In the sequel we prove that the statement is true for any $n \geq 2$ whenever it holds for $n - 1$.

Suppose that $\nabla f(x)$ is a $P_*(\kappa)$ matrix for any $x \in \Re^n$. Let us suppose that the statement of the lemma is not true, that is, there exist two points $x \neq y \in \Re^n$ such that

$$(1 + 4\kappa) \sum_{i \in I_+^f(x,y)} (x_i - y_i)(f_i(x) - f_i(y)) + \sum_{i \in I_-^f(x,y)} (x_i - y_i)(f_i(x) - f_i(y)) \leq 0. \quad (4.4)$$

We first consider the case where there exists some $i \in I$ such that $x_i = y_i$. For simplicity we can assume that $i = n$ and consider the subfunction

$$h_i(\chi_1, \dots, \chi_{n-1}) = f_i(\chi_1, \dots, \chi_{n-1}, y_n), \quad i = 1, \dots, n - 1.$$

Since $\nabla h(\chi_1, \dots, \chi_{n-1})$ is again a $P_*(\kappa, \beta)$ matrix for any $(\chi_1, \dots, \chi_{n-1}) \in \Re^{n-1}$, the induction hypothesis implies that h is a strict $P_*(\kappa)$ mapping in $\Re^{n-1}$ and therefore for any $x \neq y \in \Re^n$ with $x_n = y_n$ there holds

$$(1 + 4\kappa) \sum_{i \in I_+^f(x,y)} (x_i - y_i)(f_i(x) - f_i(y)) + \sum_{i \in I_-^f(x,y)} (x_i - y_i)(f_i(x) - f_i(y)) > 0. \quad (4.5)$$

This relation contradicts (4.4).

Thus it remains to consider the case where $x_i \neq y_i$ for all $i \in I$. For any fixed $y \in \Re^n$, let us denote by Ω_y the set given by

$$\left\{ x \mid (1 + 4\kappa) \sum_{i \in I_+^f(x,y)} (x_i - y_i)(f_i(x) - f_i(y)) \right.$$
$$\left. + \sum_{i \in I_-^f(x,y)} (x_i - y_i)(f_i(x) - f_i(y)) \leq 0, x > y \right\}.$$

We proceed to show that Ω_y is empty. Suppose to the contrary that Ω_y is nonempty. Let us consider any convergent sequence $x^k \in \Omega_y$ with $x^k \rightarrow x$. It follows readily that $x \geq y$ and

$$(1 + 4\kappa) \sum_{i \in I_+^f(x,y)} (x_i - y_i)(f_i(x) - f_i(y)) + \sum_{i \in I_-^f(x,y)} (x_i - y_i)(f_i(x) - f_i(y)) \leq 0.$$

Now we have three cases: (i) $x \neq y$ but $x_i = y_i$ for some $i \in I$; (ii) $x = y$; (iii) $x > y$. The first case (i) is impossible because otherwise from the first part of our proof we already know that inequality (4.5) holds if there are some $x_i = y_i$ and $x \neq y$. If case (ii) holds, then

$$\lim_{k \to \infty} \frac{1}{\|x^k - y\|} (f(x^k) - f(y) - \nabla f(y)(x^k - y)) = 0.$$

Let us denote $\Lambda^k = \text{diag} (x_1^k - y_1, x_2^k - y_2, ..., x_n^k - y_n)$. It follows that

$$\lim_{k \to \infty} \frac{1}{\|x^k - y\|^2} \Lambda^k (f(x^k) - f(y) - \nabla f(y)(x^k - y)) = 0.$$

Observe that the sequence $\{(x^k - y)/\|x^k - y\|\}$ is bounded and thus has at least one accumulation point. Without loss of generality, we can further assume that

$$\lim_{k \to \infty} \frac{x^k - y}{\|x^k - y\|} = u, \quad \|u\| = 1.$$

Denote

$$I_-^f(x, u) = \{i \in I : u_i(\nabla f(y)u)_i < 0\}; \quad I_+^f(x, u) = \{i \in I : u_i(\nabla f(y)u)_i \ge 0\}.$$

Then one can easily see that there exists a sufficiently large integer $\tilde{k}$ such that for any $k \ge \tilde{k}$,

$$I_-^f(x^k, y) \subseteq I_-^f(x, u); \quad I_+^f(x^k, y) \supseteq I_+^f(x, u).$$

Since $\kappa \ge 0$, we have

$$\lim_{k \to \infty} \frac{(1 + 4\kappa) \sum\limits_{i \in I_+^f(x^k, y)} (x_i^k - y_i)(f_i(x^k) - f_i(y)) + \sum\limits_{i \in I_-^f(x^k, y)} (x_i^k - y_i)(f_i(x^k) - f_i(y))}{\|x^k - y\|^2}$$

$$\ge \lim_{k \to \infty} \frac{(1 + 4\kappa) \sum\limits_{i \in I_+^f(u, y)} (x_i^k - y_i)(f_i(x^k) - f_i(y)) + \sum\limits_{i \in I_-^f(u, y)} (x_i^k - y_i)(f_i(x^k) - f_i(y))}{\|x^k - y\|^2}$$

$$= (1 + 4\kappa) \sum_{i \in I_+^f(u, y)} u_i(\nabla f(y)u)_i + \sum_{i \in I_-^f(u, y)} u_i(\nabla f(y)u)_i \ge \beta \|u\|^2 = \beta,$$

where the last inequality is implied by the assumption in the lemma that ∇f is a $P_*(\kappa, \beta)$ matrix with $\beta > 0$. The above relation implies that for sufficiently large k, the inequality

$$(1 + 4\kappa) \sum_{i \in I_+^f(x^k, y)} (x_i^k - y_i)(f_i(x^k) - f_i(y)) + \sum_{i \in I_-^f(x^k, y)} (x_i^k - y_i)(f_i(x^k) - f_i(y)) > 0$$

holds, which contradicts the assumption $x^k \in \Omega_y$. This implies that y does not belong to the boundary of Ω_y. The above discussion shows that cases (i) and (ii) are impossible. Hence only case (iii) remains. In this situation, we have $x \in \Omega_y$, which further implies Ω_y is closed. Let us define

$$u = \arg \min_{x \in \Omega_y} \|x - y\|. \tag{4.6}$$

If Ω_y is nonempty, then we know that u is (perhaps not uniquely) well defined. Moreover, for any u satisfying relation (4.6), one can easily prove the following conclusion:

$$x \in \Omega_y \text{ and } x \le u \quad \Rightarrow \quad x = u. \tag{4.7}$$

Since $\nabla f(u)$ is a $P_*(\kappa, \beta)$ matrix and thus a P matrix, by Lemma 4.2.2 there is a vector $h < 0$ such that $\nabla f(u)h < 0$. It follows immediately that

$$\lim_{t \to 0} \frac{1}{t} (f(u + th) - f(u)) = \nabla f(u)h < 0.$$

Because $y < u \in \Omega_y$, one can choose sufficiently small $t > 0$ such that the relations $u > u + th > y$ and $f(u + th) - f(u) < 0$ hold. From the continuity of $f(x)$ it follows

$$(1 + 4\kappa) \sum_{i \in I^f_+(u+th,y)} (u_i + th_i - y_i)(f_i(u + th) - f_i(y))$$

$$+ \sum_{i \in I^f_-(u+th,y)} (u_i + th_i - y_i)(f_i(u + th) - f_i(y)) < 0.$$

The above discussion means that $y < u + th < u$ and $u + th \in \Omega_y$ for sufficiently small $t > 0$, which contradicts the statement (4.7). Hence case (iii) cannot be true and this further implies that Ω_y is empty.

Now suppose that $x \ne y$ satisfies (4.4). Then $x_i \ne y_i$ for any $i \in I$; otherwise it will contradict the first part of the proof. Let

$$\Lambda := \text{diag}(\text{sign}(x_1 - y_1), \ldots, \text{sign}(x_n - y_n)), \quad \tilde{f}(x) := \Lambda f(\Lambda x).$$

Then for any $x \in \Re^n$, $\nabla \tilde{f}(x)$ is a $P_*(\kappa, \beta)$ matrix, since the diagonal matrix Λ is nonsingular. Moreover, by the construction of $\tilde{f}(x)$, the relations $\tilde{x} = \Lambda x > \Lambda y = \tilde{y}$ hold. It follows directly that

$$(1 + 4\kappa) \sum_{i \in I^{\tilde{f}}_+(\tilde{x},\tilde{y})} (\tilde{x}_i - \tilde{y}_i)(\tilde{f}_i(\tilde{x}) - \tilde{f}_i(\tilde{y})) + \sum_{i \in I^{\tilde{f}}_-(\tilde{x},\tilde{y})} (\tilde{x}_i - \tilde{y}_i)(\tilde{f}_i(\tilde{x}) - \tilde{f}_i(\tilde{y}))$$

$$= (1 + 4\kappa) \sum_{i \in I^f_+(x,y)} (x_i - y_i)(f_i(x) - f_i(y)) + \sum_{i \in I^f_-(x,y)} (x_i - y_i)(f_i(x) - f_i(y))$$

$$\le 0,$$

which is a contradiction of the second part of our proof. From our discussions above we have seen that for any $x \ne y \in \Re^n$, the inequality (4.4) does not hold. Therefore $f(x)$ is a strict $P_*(\kappa)$ mapping. This completes the proof of the lemma.   □

Now we are ready to state the major result in this section, which is a combination of Lemma 4.2.8 and Lemma 4.2.9.

Proposition 4.2.10   *Suppose that $f(x) : \Re^n \to \Re^n$ is continuously differentiable. Then $f(x)$ is a $P_*(\kappa)$ mapping if and only if $\nabla f(x)$ is a $P_*(\kappa)$ matrix for any $x \in \Re^n$.*

Proof   The necessary part of the proposition follows from Lemma 4.2.8. Hence it remains to prove the sufficient part. Since $\nabla f(x)$ is a $P_*(\kappa)$ matrix for any $x \in \Re^n$, it is trivial to see that for any $\beta > 0$ and $x \in \Re^n$ $\nabla f(x) + \beta E$ is a $P_*(\kappa, \beta)$ matrix. Therefore, by Lemma 4.2.9, we deduce that the mapping $f(x) + \beta x$ is a $P_*(\kappa)$ mapping for any $\beta > 0$. Now recalling Corollary 4.2.6, one can conclude that $f(x)$ is a $P_*(\kappa)$ mapping. $\qquad\square$

4.3 NEW SEARCH DIRECTIONS FOR $P_*(\kappa)$ CPS

In this section we introduce some new IPMs for solving $P_*(\kappa)$ CPs. The new IPMs are based on the so-called *self-regular* functions and *self-regular* proximities analogous to those we introduced early in Chapter 3 for LO. To facilitate our discussion about new IPMs, we need to introduce more notation. First we remind the reader that whenever there is no possible confusion, for convenience we use capital letters to denote the diagonal matrix obtained from a vector; for instance $D = \text{diag}(d)$. As in the LO chapter, for every $(x, s) > 0$ and $\mu > 0$, we define

$$v := \sqrt{\frac{xs}{\mu}}, \quad v_{\min} := \min\{v_i : i \in I\}, \quad v_{\max} := \max\{v_i : i \in I\}. \quad (4.8)$$

The proximity for CP is defined by

$$\Phi(x, s, \mu) := \Psi(v) = \sum_{i=1}^{n} \psi(v_i). \quad (4.9)$$

Correspondingly we say the proximity $\Psi(v)$ is *self-regular* if its kernel function $\psi(t)$ is *self-regular*. Analogous to the LO case, we denote by σ the norm of the gradient of $\Psi(v)$, that is,

$$\sigma := \|\nabla \Psi(v)\|. \quad (4.10)$$

Let $\mathcal{F}_{\text{CP}}^o$ denote the strictly feasible region of CP, that is,

$$\mathcal{F}_{\text{CP}}^o := \{(x, s) \in \Re_{++}^{2n} : s = f(x)\}.$$

The neighborhood used in our algorithm is defined by

$$\mathcal{N}(\tau, \mu) = \{(x, s) \in \mathcal{F}_{\text{CP}}^o, \Psi(x, s, \mu) \le \tau\}. \quad (4.11)$$

In the present chapter, we also make the following assumption about the nonlinear mapping f.

A.1 *There exist two constants $\mathcal{L} > 0$ and $0 < \gamma \leq 1$ such that for any* $(x, s) \in \mathcal{F}_{CP}^o$, $\Delta x \in \mathfrak{R}^n$ *and any vector* $\chi = (\chi_1, ..., \chi_n)^T \in \mathfrak{R}_+^n$ *satisfying* $\|\chi x^{-1} \Delta x\|_\infty < \gamma$, *the following inequality holds:*

$$\left\| \frac{v}{s} \left(\nabla f_1(x + \chi_1 \Delta x), ..., \nabla f_n(x + \chi_n \Delta x) - \nabla f(x) \right)^T \Delta x \right\| \leq \mathcal{L} \|\chi\|_\infty \left\| \frac{v}{s} \nabla f(x) \Delta x \right\|.$$

$$(4.12)$$

We would like to point out that the constant $\mathcal{L}$ in (4.12) might depend on the constant γ. In the special case of LCPs, (4.12) is true for any $\chi \in \mathfrak{R}^n$ and $\mathcal{L} = 0$. In this case, we can remove the requirement $\gamma \leq 1$ in Assumption A.1.

Closely related to Assumption A.1 is the following condition:

A.2 *For any* $(x, s) \in \mathcal{F}_{CP}^o$, $\Delta x \in \mathfrak{R}^n$ *and any number* $\alpha \in \mathfrak{R}_{++}$ *satisfying*

$$\alpha \|x^{-1} \Delta x\| \leq \gamma \leq 1,$$

the following relation holds:

$$\left\| \frac{v}{s} \left(\frac{f(x + \alpha \Delta x) - f(x)}{\alpha} - \nabla f(x) \Delta x \right) \right\| \leq \alpha \mathcal{L} \left\| \frac{v}{s} \nabla f(x) \Delta x \right\|. \qquad (4.13)$$

Let us speculate a bit about the relation between these two conditions. First we observe that, by Taylor's expansion, inequality Assumption A.2 is satisfied if Assumption A.1 holds. This indicates that Assumption A.2 is slightly weaker than Assumption A.1. However, Assumption A.1 is more suitable for the analysis of our new IPMs based on *self-regular* proximities, which are characterized by their second derivatives. Assumption A.1 is satisfied automatically for any LCP. Nevertheless, one can easily see that these two assumptions are essentially equivalent in $\mathfrak{R}^1$. In this case, we can simply verify that the following functions $\log x$, $-\log x$, $x \log x$, x^{a_0} and $\exp(a_0 x + b_0)$ with $a_0, b_0 \in \mathfrak{R}$ satisfy Assumption A.1.

In [49], the authors essentially assumed that relation Assumption A.2 holds for all points in a specific neighborhood of the central path.[2] Therefore, in some sense, one can claim that condition 3.2 in [49] is slightly weaker than Assumption A.2. However, the constants in Condition 3.2 of [49] might depend on the mapping $f(x)$ itself as well as the neighborhood of the central path which is in principle highly relevant to the strategies used in the algorithm. This further indicates that, in some situations, it is much more difficult to estimate these constants. However, except in some special cases such as LCPs, there is no general easy way to estimate these constants. Finally we also note that relation (4.12) can be viewed as a kind of relative Lipschitz condition for the Jacobian $\nabla f(x)$.

---

[2] Note that equation (12) in Condition 3.2 of [49] includes a typo. The factor θ in the right-hand side should be θ^2 to make equation (13) is correct.

We proceed to describe the new algorithm for CPs, which is an extension of the large-update primal-dual algorithm for LO proposed in the previous chapter. Starting from a strictly feasible point, the algorithm generates a sequence in the neighborhood $\mathcal{N}(\tau, \mu)$. Thus, at each iteration, we check whether the iterate is in the neighborhood $\mathcal{N}(\tau, \mu)$. If the answer is "no", then we solve the Newton-type system

$$-\nabla f(x)\Delta x + \Delta s = 0 \qquad (4.14)$$

$$s\Delta x + x\Delta s = -\mu v \nabla \Psi(v) \qquad (4.15)$$

to get a new search direction. Since the matrix $H = -\nabla f(x)$ is a $P_*(\kappa)$ matrix, it is guaranteed that the system has a unique solution for every $\nabla\Psi(v)$ (see, e.g., [58]). For the displacement Δx, let us define

$$x(\alpha) := x + \alpha\Delta x, \quad s(\alpha) := f(x + \alpha\Delta x).$$

As we see later, by progressing properly along this search direction we will be able to reduce the value of the proximity. This procedure is repeated until the iterate enters the neighborhood $\mathcal{N}(\tau, \mu)$ again. If the present iterate is in $\mathcal{N}(\tau, \mu)$, then one reduces the barrier parameter μ by a constant ratio. The above process is repeated until the iterate is in the neighborhood and the parameter μ becomes sufficiently small. The general procedure of our algorithm can be described as follows.

Large-Update Algorithm for CP

Inputs
   A proximity parameter $\tau \geq v_1^{-1}$;
   an accuracy parameter $\varepsilon > 0$;
   a fixed barrier update parameter θ, $0 < \theta < 1$;
   (x^0, s^0) and $\mu^0 = 1$ such that $\Phi(x^0, s^0, \mu^0) \leq \tau$.
begin
   $x := x^0$; $s := s^0$; $\mu := \mu^0$;
   while $n\mu \geq \varepsilon$ **do**
   begin
      $\mu := (1 - \theta)\mu$;
      while $\Phi(x, s, \mu) > \tau$ **do**
      begin
         Solve system (4.14)–(4.15) for $\Delta x, \Delta s$,
         Determine a step size α;
         $x := x(\alpha)$;
         $s := s(\alpha)$;
      end
   end
end

Remark 4.3.1 *Note that if a strictly feasible starting point is available, then we can choose* $\tau = \max\{\Phi(x^0, s^0, \mu^0), 1/\nu_1\} \leq \mathcal{O}(n)$. *This implies the starting point is in the neighborhood* $\mathcal{N}(\tau, \mu^0)$. *Analogous to the LO case, we always assume that* $v_{max} > 1$ *in the algorithm.*

Remark 4.3.2 *Similar to the LO case, we stipulate that the step size* α *be taken so that the proximity measure function* Ψ *decreases sufficiently. A default bound for such a step size* α *is given later by (4.44).*

4.4 COMPLEXITY OF THE ALGORITHM

This section is devoted to estimating the complexity of the algorithm. The section consists of three parts. In the first subsection, we present some bounds for the norm of the search direction and the maximal feasible step size. In the second subsection we estimate the decrease of the proximity for a feasible step size. Finally, we summarize the complexity of the algorithm in the last subsection.

4.4.1 Ingredients for Estimating the Proximity

In this section, we provide certain ingredients that are used for estimating the proximity. We start by introducing some notation. For each $\alpha > 0$ and $\chi \in \Re^n$, let us define

$$\Delta s(\alpha) := \frac{1}{\alpha}(f(x + \alpha \Delta x) - f(x)), \tag{4.17}$$

$$d_x = \frac{v}{x}\Delta x, \quad d_s := \frac{v}{s}\Delta s, \tag{4.18}$$

$$d_s(\alpha) = \frac{v}{s}\Delta s(\alpha), \tag{4.19}$$

$$\nabla d_s(\alpha) = \frac{1}{\alpha}\left(\frac{v}{s}\nabla f(x + \alpha \Delta x)\Delta x\right), \tag{4.20}$$

$$\nabla d_s(\chi) = ([\nabla d_s(\chi_1)]_1, [\nabla d_s(\chi_2)]_2, ..., [\nabla d_s(\chi_n)]_n)^{\mathrm{T}} \tag{4.21}$$

$$\text{or equivalently } [\nabla d_s(\chi)]_i := \frac{1}{\chi_i}\left(\frac{v_i}{s_i}\nabla f_i(x + \chi_i \Delta x)\Delta x\right).$$

Note that the functions $\Delta s(\alpha)$ and $d_s(\alpha)$ are not defined at $\alpha = 0$. However, one can easily see that these two definitions can be extended to the case $\alpha = 0$ as

$$\Delta s(0) := \lim_{\alpha \to 0} \Delta s(\alpha) = \nabla f(x)\Delta x; \quad d_s(0) := \lim_{\alpha \to 0} d_s(\alpha) = \frac{v}{s}\nabla f(x)\Delta x. \quad (4.22)$$

It should be noted that, by using the notation introduced by (4.18), we can rewrite the system (4.14)–(4.15) as

$$-\Lambda_f d_x + d_s = 0, \quad d_x + d_s = -\nabla \Psi(v), \quad (4.23)$$

where $\Lambda_f := \mu VS^{-1}\nabla f(x)VS^{-1}$. We now present some estimates of the displacements d_x and d_s in term of σ. For this we also need the following lemma about $P_*(\kappa)$ matrices which is exactly the same as Lemma 3.4 in [58]. However we copy it here for ease of reference.

Lemma 4.4.1 *A matrix M is a $P_*(\kappa)$ matrix if and only if for any positive definite diagonal matrix Λ and any $\Delta x, \Delta s, h \in \Re^n$, the relations*

$$\Lambda^{-1}\Delta x + \Lambda \Delta s = h, \quad \Delta s = M\Delta x$$

always imply

$$\Delta x^T \Delta s \geq -\kappa \|h\|^2.$$

Now let us recall definition (4.10) of σ. By using the above lemma and following an analogous discussion to that in the proof of Lemma 3.1 in [49], one can readily obtain the following results, which present some bounds for the search direction in various spaces.

Lemma 4.4.2 *Let $(\Delta x, \Delta s)$ be the unique solution of system (4.14)–(4.15) and (d_x, d_s) be the corresponding solution of system (4.23) in the scaled v-space. Then we have*

(i) $-\kappa \sigma^2 \leq \Delta x^T \Delta s/\mu = d_x^T d_s \leq \sigma^2/4$,

(ii) $\|d_x d_s\|_\infty = \|\Delta x \Delta s\|_\infty/\mu \leq (1+\kappa)\sigma^2/4$,

(iii) $\|d_x\|^2 + \|d_s\|^2 = \|d_x + d_s\|^2 - 2d_x^T d_s \leq (1+2\kappa)\sigma^2$,

(iv) $\|x^{-1}\Delta x\| = \|v^{-1}d_x\| \leq \|d_x\|/v_{min} \leq \sigma\sqrt{1+2\kappa}/v_{min}$,

(v) $\|s^{-1}\Delta s\| = \|v^{-1}d_s\| \leq \|d_s\|/v_{min} \leq \sigma\sqrt{1+2\kappa}/v_{min}$.

Let us define

$$\hat{\alpha} := \min\left\{1, \frac{\gamma v_{min}}{\sigma\sqrt{1+2\kappa}}\right\}. \quad (4.24)$$

It follows from result (iv) of Lemma 4.4.2 that $x + \alpha \Delta x > 0$ for all $\alpha \in [0, \hat{\alpha})$. In light of the definition of $\nabla d_s(\alpha)$ in (4.20), $\nabla d_s(\alpha)$ can be represented by

$$\nabla d_s(\alpha) = \frac{1}{\alpha} \frac{\partial}{\partial \alpha} (\alpha d_s(\alpha)) \tag{4.25}$$

and equation (4.12) in Assumption A.1 can be expressed as

$$\|\chi \nabla d_s(\chi) - d_s\| \leq \mathcal{L} \|\chi\|_\infty \|d_s\|. \tag{4.26}$$

The following result follows directly from the above observations.

Lemma 4.4.3 *Suppose that Assumption A.1 holds and let $d_s(x) = \chi^{-1}(f(x + \chi \Delta x) - f(x))$. Then*

(i) $\|\chi \nabla d_s(\chi)\| \leq (1 + \alpha \mathcal{L}) \|d_s\| \leq (1 + \mathcal{L}) \|d_s\|$,

(ii) $\|\chi d_s(\chi)\| \leq \alpha(1 + \alpha \mathcal{L}) \|d_s\| \leq \alpha(1 + \mathcal{L}) \|d_s\|$ *for every $\alpha \in [0, \hat{\alpha})$ and every vector χ satisfying $0 \leq \chi \leq \alpha e$.*

Proof Assertion (i) follows directly from (4.26) and the relations $\|\chi\|_\infty \leq \alpha$ and $\alpha \leq \hat{\alpha}$. To confirm Assertion (ii), we observe that for any $i \in I$, by using Taylor's series expansion, one can deduce

$$[\chi d_s(\chi)]_i = \frac{v_i}{s_i} (f_i(x + \chi_i \Delta x) - f_i(x))$$

$$= \frac{v_i}{s_i} (f_i(x) + \chi_i \nabla f_i(x + \chi_i' \Delta x) \Delta x - f_i(x))$$

$$= \frac{v_i}{s_i} \chi_i \nabla f_i(x + \chi_i' \Delta x) \Delta x$$

$$= \chi_i [\chi' \nabla d_s(\chi')]_i,$$

where $\chi_i' \in (0, \chi_i)$. Since $\chi_i' < \chi_i \leq \alpha$ for all $i \in I$, it follows from (i) that

$$\|\chi d_s(\chi)\| = \|\chi(\chi' \nabla d_s(\chi'))\| \leq \|\chi\|_\infty \|\chi' \nabla d_s(\chi')\|$$

$$\leq \alpha \|\chi' \nabla d_s(\chi')\| \leq \alpha(1 + \alpha \mathcal{L}) \|d_s\|.$$

This completes the proof of Assertion (ii). $\qquad \square$

Note that the constant $\hat{\alpha}$ has already provided a lower bound for a step size to keep the feasibility of $x(\alpha) = x + \alpha \Delta x$. However, we do not know whether for all $\alpha \in (0, \hat{\alpha})$, the displacement $s(\alpha)$ is also strictly feasible. In what follows we estimate the growth behavior of the norm of $s(\alpha)$ for all $\alpha \in (0, \hat{\alpha})$. This further gives a lower bound for a strictly feasible step size for both $x(\alpha)$ and $s(\alpha)$. By combining Lemma 4.4.2, Lemma 4.4.3 and Proposition 3.1.5, we obtain the following result.

Lemma 4.4.4 *Suppose that Assumption A.1 holds and that the function* $\Psi(v)$ *given by (4.9) is self-regular. Then for any* $\alpha \in (0, \hat{\alpha})$,

$$\|(x^{-1}\Delta x, s^{-1}\Delta s(\alpha))\| = \|(v^{-1}d_x, v^{-1}d_s(\alpha))\| \leq \bar{\alpha}^{-1} \leq v_5\sigma\left(1 + \frac{q\sigma}{v_1}\right)^{1/q},$$

where

$$\bar{\alpha} := \frac{\hat{\alpha}}{(1 + \mathcal{L})} \tag{4.27}$$

and

$$v_5 := \sqrt{1 + 2\kappa}(1 + \mathcal{L}). \tag{4.28}$$

Furthermore, the maximal step size satisfies $\alpha_{\max} \geq \bar{\alpha}$.

Proof By the definition (4.18), we see that

$$\|(x^{-1}\Delta x, s^{-1}\Delta s)\| = \|(v^{-1}d_x, v^{-1}d_s)\|$$

$$\leq \frac{1}{v_{\min}}\sqrt{\|d_x\|^2 + \|d_s\|^2}$$

$$\leq \frac{\sigma\sqrt{1 + 2\kappa}}{v_{\min}}$$

$$\leq \sigma\left(1 + \frac{q\sigma}{v_1}\right)^{1/q}\sqrt{1 + 2\kappa}$$

where the second and third inequalities follow from (iii) of Lemma 4.4.2 and (3.11) in Proposition 3.1.5, respectively. Since

$$\|d_s(\alpha)\| \leq (1 + \mathcal{L})\|d_s\|$$

for every $\alpha \in (0, \hat{\alpha})$, from result (ii) of Lemma 4.4.3 and Proposition 3.1.5 it follows that

$$\|(x^{-1}\Delta x, s^{-1}\Delta s(\alpha))\| = \|(v^{-1}d_x, v^{-1}d_s(\alpha))\|$$

$$\leq \frac{1}{v_{\min}}\sqrt{\|d_x\|^2 + \|d_s(\alpha)\|^2}$$

$$\leq \frac{1}{v_{min}} \sqrt{\|d_x\|^2 + (1+\mathcal{L})^2 \|d_s\|^2}$$

$$\leq \frac{1}{v_{min}} (1+\mathcal{L}) \sqrt{\|d_x\|^2 + \|d_s\|^2}$$

$$\leq \frac{v_5 \sigma}{v_{min}}$$

$$\leq v_5 \sigma \left(1 + \frac{q\sigma}{v_1}\right)^{1/q}$$

for every $\alpha \in (0, \hat{\alpha})$. Note that a step size α is feasible if and only if both $x + \alpha \Delta x \geq 0$ and $s + \alpha \Delta s(\alpha) \geq 0$. This gives the last statement of the lemma. $\square$

We remark that since $\mathcal{L} \geq 0$, obviously $\bar{\alpha} \leq \hat{\alpha}$.

4.4.2 Estimate of the Proximity After a Step

We are going to estimate the decrease of the proximity for a feasible step size. First let us define

$$v(\alpha) = \sqrt{\frac{x(\alpha)s(\alpha)}{\mu}} = \sqrt{(v + \alpha d_x)(v + \alpha d_s(\alpha))}$$

$$= v \sqrt{(e + \alpha v^{-1} d_x)(e + \alpha v^{-1} d_s(\alpha))}. \qquad (4.29)$$

Since the proximity after one feasible step is defined by $\Psi(v(\alpha))$, to estimate the decrease of the proximity for a step size α, it suffices to consider the gap of the proximities before and after one step, which is defined as a function of the step size α:

$$g(\alpha) := \Psi(v(\alpha)) - \Psi(v). \qquad (4.30)$$

Because the function $\Psi(v)$ is *self-regular*, from condition SR2 it follows directly that

$$g(\alpha) = \sum_{i=1}^{n} (\psi(v_i(\alpha)) - \psi(v_i))$$

$$\leq -\Psi(v) + \frac{1}{2} \sum_{i=1}^{n} (\psi(v_i + \alpha[d_x]_i) + \psi(v_i + \alpha[d_s(\alpha)]_i).$$

For any $i \in I$, let us define

$$\varphi_i(\alpha) := \psi(v_i + \alpha[d_s(\alpha)]_i).$$

Obviously

$$\varphi_i(\alpha) = \varphi_i(0) + \alpha\varphi_i'(0) + \int_0^\alpha \left(\varphi_i'(\xi) - \varphi_i'(0)\right)d\xi.$$

Moreover, by simple calculus and using the notation defined by (4.19), (4.20) and the relation (4.22), one can readily check that the terms in the previous equation can be written as

$$\varphi_i(0) = \psi(v_i),$$

$$\varphi_i'(\alpha) = \psi'(v_i + \alpha[d_s(\alpha)]_i)[\alpha\nabla d_s(\alpha)]_i$$

$$= \psi'(v_i + \alpha[d_s(\alpha)]_i)\left(\frac{v}{s}\nabla f(x + \alpha\Delta x)\Delta x\right)_i, \quad (4.31)$$

$$\varphi_i'(0) = \psi'(v_i)\left(\frac{v}{s}\nabla f(x)\Delta x\right)_i$$

$$= \psi'(v_i)[d_s]_i. \quad (4.32)$$

Thus we obtain that

$$\psi(v_i + \alpha[d_s(\alpha)]_i) = \psi(v_i) + \alpha\psi'(v_i)[d_s]_i + \int_0^\alpha (\varphi_i'(\xi) - \varphi_i'(0))d\xi.$$

Similarly, we have

$$\psi(v_i + \alpha[d_x]_i) = \psi(v_i) + \alpha\psi'(v_i)[d_x]_i + \int_0^\alpha (r_i'(\xi) - r_i'(0))d\xi,$$

where

$$r_i(\alpha) := \psi(v_i + \alpha[d_x]_i).$$

Therefore, from (4.9), (4.10) and (4.23), we conclude that the summation is given by

$$\sum_{i=1}^n \left(\psi(v_i + \alpha[d_x]_i) + \psi(v_i + \alpha[d_s(\alpha)]_i)\right)$$

$$= 2\sum_{i=1}^n \psi(v_i) + \alpha\sum_{i=1}^n \psi'(v_i)[d_x + d_s]_i + g_1(\alpha)$$

$$= 2\Psi(v) - \alpha\|\nabla\Psi(v)\|^2 + g_1(\alpha)$$

$$= 2\Psi(v) - \alpha\sigma^2 + g_1(\alpha), \quad (4.33)$$

where

$$g_1(\alpha) := \sum_{i=1}^{n} \int_{0}^{\alpha} (r_i'(\xi) - r_i'(0)) d\xi + \sum_{i=1}^{n} \int_{0}^{\alpha} (\varphi_i'(\xi) - \varphi_i'(0)) d\xi$$

$$= \int_{0}^{\alpha} \left(\sum_{i=1}^{n} (r_i'(\xi) - r_i'(0)) + \sum_{i=1}^{n} (\varphi_i'(\xi) - \varphi_i'(0)) \right) d\xi.$$

As we have done in the LO case, to get a good estimation of the value $g_1(\alpha)$, it is essential to estimate first the derivatives $r_i'(\alpha)$ and $\varphi_i'(\alpha)$. This is the major task of the following lemma.

Lemma 4.4.5 *Suppose that Assumption A.1 holds and that the proximity $\Psi(v)$ is self-regular. Then for every $\alpha \in (0, \bar{\alpha})$, we have*

(i)

$$\max_{i \in I} \{ \psi''(v_i + \alpha[d_x]_i), \psi''(v_i + \alpha[d_s(\alpha)]_i) \} \leq \nu_2 \omega(\alpha),$$

where

$$\omega(\alpha) := (v_{\max} + \alpha \nu_5 \sigma)^{p-1} + (v_{\min} - \alpha \nu_5 \sigma)^{-q-1}; \qquad (4.34)$$

(ii)

$$\sum_{i=1}^{n} (r_i'(\alpha) - r_i'(0)) + \sum_{i=1}^{n} (\varphi_i'(\alpha) - \varphi_i'(0)) \leq \alpha \nu_2 \nu_6 \sigma^2 \omega(\alpha),$$

where

$$\nu_6 := \nu_5^2 + \frac{L\sqrt{1 + 2\kappa}}{\nu_2}. \qquad (4.35)$$

Proof We first prove inequality (i). By Lemma 4.4.4 and the assumption on α we know that the step size in the lemma satisfies $(v + \alpha[d_x], v + \alpha[d_s(\alpha)]) > 0$. Since the proximity $\Psi(v)$ is *self-regular*, from condition SR1 we obtain

$$\psi''(v_i + \alpha[d_x]_i) \leq \nu_2 \left((v_i + \alpha[d_x]_i)^{p-1} + (v_i + \alpha[d_x]_i)^{-q-1} \right),$$

$$\psi''(v_i + \alpha[d_s(\alpha)]_i) \leq \nu_2 \left((v_i + \alpha[d_s(\alpha)]_i)^{p-1} + (v_i + \alpha[d_s(\alpha)]_i)^{-q-1} \right).$$

Note that result (iii) of Lemma 4.4.2 and result (ii) of Lemma 4.4.3 ensure that

$$\alpha \|d_x\| \leq \alpha \sigma \sqrt{1 + 2\kappa}, \quad \alpha \|d_s(\alpha)\| \leq \alpha \nu_5 \sigma.$$

Combination of the above relations leads to the desired Assertion (i).

We proceed to consider Assertion (ii). For this we first prove the following inequality:

$$\sum_{i=1}^{n}(r'_i(\alpha) - r'_i(0) + \varphi'_i(\alpha) - \varphi'_i(0)) \le \alpha\Big\{\nu_2\nu_5^2\omega(\alpha) + \sqrt{1 + 2\kappa\mathcal{L}}\Big\}\sigma^2. \quad (4.36)$$

By using (4.31), (4.32) and the Mean-Value Theorem [100], we obtain

$$\sum_{i=1}^{n}(\varphi'_i(\alpha) - \varphi'_i(0)) = \sum_{i=1}^{n}\{(\psi'(v_i + \alpha[d_s(\alpha)]_i)(\alpha[\nabla d_s(\alpha)]_i) - \psi'(v_i)[d_s]_i)\}$$

$$= \sum_{i=1}^{n}\{\alpha(\psi'(v_i + \alpha[d_s(\alpha)]_i) - \psi'(v_i))[\nabla d_s(\alpha)]_i + \psi'(v_i)(\alpha[\nabla d_s(\alpha)]_i - [d_s]_i)\}$$

$$= \sum_{i=1}^{n}\Big\{\alpha^2\psi''(v_i + \overline{\chi}_i[d_s(\overline{\chi}_i)]_i)\chi_i[\nabla d_s(\chi_i)]_i[\nabla d_s(\alpha)]_i + \psi'(v_i)(\alpha[\nabla d_s(\alpha)]_i - [d_s]_i)\Big\}$$

for some $0 \le \chi = (\chi_1, \chi_2, \ldots, \chi_n)^{\mathrm{T}}$, $\overline{\chi} = (\overline{\chi}_1, \ldots, \overline{\chi}_n)^{\mathrm{T}} \le \alpha e$. Now, by making use of Assertion (i) and applying the Cauchy–Schwarz inequality to the vectors $\psi'(v)$ and $\alpha\nabla d_s(\alpha) - d_s$, we obtain

$$\sum_{i=1}^{n}\Big\{\alpha^2\psi''(v_i + \overline{\chi}_i[d_s(\overline{\chi}_i)]_i)\chi_i[\nabla d_s(\chi_i)]_i[\nabla d_s(\alpha)]_i + \psi'(v_i)(\alpha[\nabla d_s(\alpha)]_i - [d_s]_i)\Big\}$$

$$\le \alpha^2\nu_2\omega(\alpha)\sum_{i=1}^{n}|\chi_i[\nabla d_s(\chi_i)]_i|\cdot|[\nabla d_s(\alpha)]_i| + \sum_{i=1}^{n}|\psi'(v_i)|\cdot|\alpha[\nabla d_s(\alpha)]_i - [d_s]_i|$$

$$\le \alpha\nu_2\omega(\alpha)\|\chi[\nabla d_s(\chi)]\|\cdot\|\alpha[\nabla d_s(\alpha)]\| + \|\nabla\Psi(v)\|\cdot\|\alpha\nabla d_s(\alpha) - d_s\|. \quad (4.37)$$

From Assumption A.1 and Lemma 4.4.3 we conclude that

$$\|\chi[\nabla d_s(\chi)]\|\cdot\|\alpha[\nabla d_s(\alpha)]\| \le (1 + \mathcal{L})^2\|d_s\|^2$$

and

$$\|\nabla\Psi(v)\|\cdot\|\alpha\nabla d_s(\alpha) - d_s\| \le \sigma\cdot\alpha\mathcal{L}\|d_s\|.$$

These two inequalities combined with (4.37) further imply

$$\sum_{i=1}^{n}(\varphi'_i(\alpha) - \varphi'_i(0)) \le \alpha\nu_2(1 + \mathcal{L})^2\omega(\alpha)\|d_s\|^2 + \sigma\cdot\alpha\mathcal{L}\|d_s\|. \quad (4.38)$$

Similarly, we can prove that

$$\sum_{i=1}^{n}(r'_i(\alpha) - r'_i(0)) \le \alpha\nu_2\omega(\alpha)\|d_x\|^2. \quad (4.39)$$

Recalling the inequalities

$$\|d_x\|^2 + \|d_s\|^2 \le (1 + 2\kappa)\sigma^2,$$

$$\|d_s\| \le \sqrt{1 + 2\kappa}\,\sigma,$$

from Lemma 4.4.2 and substituting them into (4.38) and (4.39), we readily obtain

$$\sum_{i=1}^{n}(r'_i(\alpha) - r'_i(0) + \varphi'_i(\alpha) - \varphi'_i(0))$$

$$\le \alpha\nu_2\omega(\alpha)\|d_x\|^2 + \alpha\nu_2(1 + \mathcal{L})^2\omega(\alpha)\|d_s\|^2 + \sigma\cdot\alpha\mathcal{L}\|d_s\|$$

$$\le \alpha\nu_2\nu_5^2\omega(\alpha)\sigma^2 + \alpha\mathcal{L}\sqrt{1 + 2\kappa}\,\sigma^2,$$

which gives (4.36). Now let us recall that $\omega(\alpha) \ge 1$ whenever $v_{\max} \ge 1$. It follows directly from (4.36) that

$$\sum_{i=1}^{n}(r'_i(\alpha) - r'_i(0)) + \sum_{i=1}^{n}(\varphi'_i(\alpha) - \varphi'_i(0)) \le \alpha\nu_2\left(\nu_5^2 + \frac{\sqrt{1 + 2\kappa}\,\mathcal{L}}{\nu_2}\right)\omega(\alpha)\sigma^2,$$

(4.40)

which gives Assertion (ii). $\square$

We progress to discuss the decreasing behavior of $\Psi(\alpha)$ for a strictly feasible step size α. By (4.33) and Lemma 4.5, we see that

$$-\sigma^2\alpha + g_1(\alpha) \le -\sigma^2\alpha + \nu_2\nu_6\sigma^2\int_0^\alpha \zeta\omega(\zeta)d\zeta := g_2(\alpha). \qquad (4.41)$$

The function $g_2(\alpha)$ is strictly convex and twice differentiable for $\alpha \in [0, \overline{\alpha})$. Further it is decreasing at $\alpha = 0$ and its derivative increases as α goes to $\overline{\alpha}$. Therefore, it attains its global minimum either at the extreme point $\overline{\alpha}$ or at its unique stationary point α^*, which is the unique solution of the system

$$0 = -1 + \nu_2\nu_4\alpha\big((v_{\max} + \alpha\nu_3\sigma)^p + (v_{\min} - \alpha\nu_3\sigma)^{-q}\big). \qquad (4.42)$$

There are two cases for the solution α^* of the above system. The first case is $\alpha^* \ge \overline{\alpha}$. In this situation we immediately obtain a lower bound for α^* from Lemma 4.4.4. In the sequel we focus only on the second case and discuss how to estimate the value of α^* when $\alpha^* < \overline{\alpha}$. For this we need some technical results about the roots of certain equations. These conclusions have a key role in our later estimation about the stationary point of $g_2(\alpha)$.

Lemma 4.4.6 *Let $p > 1$ and $\rho > 0$ be two given constants.*

(i) *If t_* is the unique solution of the equation $t(1 + t)^{p-1} - \rho = 0$, then*

$$t_* \geq \frac{\rho}{1 + (p - 1)\rho}.$$

(ii) *If $t_* \in (0, 1)$ is the unique solution of the equation $t(1 - t)^{-1-p} = \rho$, then*

$$t_* \geq \frac{\rho}{1 + \rho(p + 1)}.$$

Proof First we note that $t_* > 0$ in view of the assumption of the lemma. Hence for case (i),

$$(1 + t_*)^p = \rho\left(1 + \frac{1}{t_*}\right) > 1.$$

It follows that

$$t_* = \left(\rho + \frac{\rho}{t_*}\right)^{1/p} - 1 = \left(1 - \frac{\rho + \rho t_* - t_*}{\rho + \rho t_*}\right)^{-1/p} - 1$$

$$\geq \left(1 - \frac{\rho + \rho t_* - t_*}{p\rho + p\rho t_*}\right)^{-1} - 1 = \frac{\rho + \rho t_* - t_*}{(p - 1)(\rho + \rho t_*) + t_*} > 0, \qquad (4.43)$$

where the inequality is given by (1.10) and $\rho + \rho t_* - t_* > 0$. Let us define

$$h(t) = (1 + \rho(p - 1))t^2 + (1 + \rho(p - 1) - \rho)t - \rho.$$

One can easily verify that the equation system $h(t) = 0$ has two roots

$$t_1 = -1; \quad t_2 = \frac{\rho}{1 + \rho(p - 1)}.$$

Note that from (4.43) we obtain $h(t_*) > 0$. Since $t_* > 0$, the above relation means immediately

$$t_* \geq t_2 = \frac{\rho}{1 + \rho(p - 1)},$$

as required.

Now let us switch to case (ii). If we define $\bar{t} = t_*/(1 - t_*)$, then the equation in (ii) can be written as

$$\bar{t}(1 + \bar{t})^p = \rho.$$

It follows directly from case (i) of the lemma that

$$\bar{t} \geq \frac{\rho}{1 + p\rho},$$

which further implies

$$t_* \geq \frac{\rho}{1 + \rho(p+1)}.$$

The proof of the lemma is completed.   □

We are now ready to present one of our major results in this section. A lower bound for the global minimizer α^* of $g_2(\alpha)$ is obtained as the unique solution of equation (4.42).

Lemma 4.4.7 *Let α^* be the solution of (4.42). Suppose that $\Psi(v) \geq v_1^{-1}$ and $v_{\max} > 1$. Then*

$$\alpha* \geq \gamma v_7 \sigma^{-(q+1)/q}, \tag{4.44}$$

where γ is the constant in Assumption A.1,

$$v_7 := \min\left\{\frac{v_1}{2v_2 v_6(v_1+p) + v_1 v_5(p-1)}, \right.$$

$$\left. \frac{v_1^2}{(1+v_1)(2v_2 v_6(v_1+q) + v_1 v_5(q+1))}\right\}, \tag{4.45}$$

and v_5 and v_6 are defined by (4.27) and (4.28) respectively.
  In some special cases, the constant v_7 can be simplified as follows:

  (i) If $\kappa = \mathcal{L} = 0$, that is, the CP is linear and monotone, then $v_6 = v_5 = 1$ and

$$v_7 = \min\left\{\frac{v_1}{2v_2(p+v_1) + v_1(p-1)}, \frac{v_1^2}{(1+v_1)(2v_2(q+v_1) + v_1(q+1))}\right\}.$$

  (ii) If the proximity $\psi(v)$ used in the algorithm is defined by the function $\psi(t) = \Upsilon_{p,q}(t)$ with $v_1 = v_2 = 1$, then

$$v_7 = \min\left\{\frac{1}{2v_6(p+1) + v_5(p-1)}, \frac{1}{2(v_5 + 2v_6)(q+1)}\right\}.$$

  (iii) Under both of the assumptions in (i) and (ii),

$$v_7 = \min\left\{\frac{1}{3p+1}, \frac{1}{6q+6}\right\}.$$

Proof We first mention that the lemma holds trivially if the solution α^* of the equation system (4.42) satisfies $\alpha^* \geq \bar{\alpha}$. Thus it suffices to consider the case

$\alpha^* < \overline{\alpha}$. In such a situation, let us define

$$\omega_1(\alpha) = -\frac{1}{2} + \nu_2\nu_6\alpha(\nu_{max} + \alpha\nu_5\sigma)^{p-1}$$

and

$$\omega_2(\alpha) = -\frac{1}{2} + \nu_2\nu_6\alpha(\nu_{min} - \alpha\nu_5\sigma)^{-q-1}.$$

It is easy to see that both $\omega_1(\alpha)$ and $\omega_2(\alpha)$ are increasing functions of α for $\alpha \in [0, \overline{\alpha})$. Let us further denote by α_1^* and α_2^* the roots of the equation systems $\omega_1(\alpha) = 0$ and $\omega_2(\alpha) = 0$, respectively. One can readily verify that the relation $\omega_1(\alpha_1^*) = 0$ can be equivalently written as

$$t_1(1 + t_1)^{p-1} = \rho_1,$$

where

$$t_1 = \alpha_1^* \nu_5 \nu_{max}^{-1} \sigma, \quad \rho_1 = \frac{\nu_5\sigma}{2\nu_2\nu_6\nu_{max}^p}.$$

It follows from statement (i) of Lemma 4.4.6 that

$$t_1 \geq \frac{\rho_1}{1 + \rho_1(p-1)},$$

which immediately yields

$$\alpha_1^* \geq \frac{\rho_1\nu_{max}\nu_5^{-1}\sigma^{-1}}{1 + \rho_1(p-1)} = \frac{\nu_{max}}{2\nu_2\nu_6\nu_{max}^p + (p-1)\nu_5\sigma}$$

$$\geq \frac{1}{2\nu_2\nu_6\nu_{max}^p + (p-1)\nu_5\sigma} \geq \frac{\nu_1}{2\nu_2\nu_6(\nu_1 + p\sigma) + \nu_1\nu_5(p-1)\sigma}$$

$$\geq \frac{\nu_1}{2\nu_2\nu_6(\nu_1 + p) + \nu_1\nu_5(p-1)}\sigma^{-1},$$

where the second inequality is implied by $\nu_{max} \geq 1$, the third inequality is given by (3.12), and the last inequality is true because the assumption $\Psi(\nu) \geq \nu_1^{-1}$ in the lemma implies $\sigma \geq 1$.

We next estimate the root α_2^* of equation $w_2(\alpha) = 0$. Note that we can rewrite the equation $\omega(\alpha_2^*) = 0$ as

$$t_2(1 - t_2)^{-q-1} = \rho_2,$$

where

$$t_2 = \alpha_2^* \nu_5 \nu_{min}^{-1} \sigma, \quad \rho_2 = \frac{\nu_5\nu_{min}^q\sigma}{2\nu_2\nu_6}.$$

From statement (ii) of Lemma 4.4.6 we can conclude

$$t_2 \geq \frac{p_2}{1 + p_2(q + 1)},$$

which further gives

$$\alpha_2^* \geq \frac{p_2 v_5^{-1} v_{\min} \sigma^{-1}}{1 + p_2(q + 1)} = \frac{v_{\min}^q}{2v_2 v_6 + v_5 \sigma v_{\min}^q(q + 1)} v_{\min}$$

$$\geq \frac{v_1}{2v_2 v_6(v_1 + q\sigma) + \sigma v_1 v_5(q + 1)} v_{\min}, \qquad (4.46)$$

where the last inequality follows from (3.11) and the fact that the function

$$\frac{t}{2v_2 v_6 + v_5(q + 1)t}$$

is increasing with respect to $t > 0$. Further, since $\sigma \geq 1$, by making use of (3.11) and (1.10) we obtain

$$v_{\min} \geq \left(1 + \frac{q\sigma}{v_1}\right)^{-1/q} \geq \left(1 + \frac{q}{v_1}\right)^{-1/q} \sigma^{-1/q} \geq \frac{v_1}{1 + v_1} \sigma^{-1/q}.$$

This relation, along with (4.46), yields

$$\alpha_2^* \geq \frac{v_1^2}{(1 + v_1)(2v_2 v_6(v_1 + q\sigma) + \sigma v_1 v_5(q + 1))} \sigma^{-1/q}$$

$$\geq \frac{v_1^2}{(1 + v_1)(2v_2 v_6(v_1 + q) + v_1 v_5(q + 1))} \sigma^{-(q+1)/q},$$

where the last inequality is implied by the fact that $\sigma \geq 1$.

Since equation (4.42) is equivalent to

$$\omega_1(\alpha^*) + \omega_2(\alpha^*) = 0,$$

α^* should satisfy

$$\alpha^* \geq \min\{\alpha_1^*, \alpha_2^*\},$$

and thus relation (4.44) follows immediately from the fact that $\sigma \geq 1$. By specifying the parameter v_1, v_2, v_5 and v_6 in various special cases, we easily obtain results (i)–(iii). $\qquad \square$

By a discussion similar to the proof of Theorem 3.3.4, we obtain the following.

Theorem 4.4.8 *Let the function $g(\alpha)$ be defined by (4.30) with $\Psi(v) \geq v_1^{-1}$. Then the step size α^* defined by (4.44) is strictly feasible. Moreover,*

$$g(\alpha^*) \le -\frac{\gamma \nu_7}{4} \sigma^{(q-1)/(2q)} \le -\frac{\gamma \nu_7 \nu_1^{(q-1)/(2q)}}{4} \Psi(v)^{(q-1)/(2q)}.$$

Here ν_7 is defined by (4.45). Some special cases are as follows:

(i) If $\kappa = \mathcal{L} = 0$, that is, the CP is linear and monotone and $\gamma = 1$, then $\nu_6 = \nu_5 = 1$ and

$$g(\alpha^*) \le -\frac{1}{4} \min\left\{\frac{\nu_1 \sigma^{(q-1)/(2q)}}{2\nu_2(p + \nu_1) + \nu_1(p - 1)}, \frac{\nu_1^2 \sigma^{(q-1)/(2q)}}{(1 + \nu_1)(2\nu_2(q + \nu_1) + \nu_1(q + 1))}\right\}.$$

(ii) If the proximity $\psi(v)$ used in the algorithm is defined by the function $\psi(t) = \Upsilon_{p,q}(t)$ with $\nu_1 = \nu_2 = 1$, then

$$g(\alpha^*) \le -\frac{\gamma}{4} \min\left\{\frac{\sigma^{(q-1)/(2q)}}{2\nu_6(p + 1) + \nu_5(p - 1)}, \frac{\sigma^{(q-1)/(2q)}}{2(\nu_5 + 2\nu_6)(q + 1)}\right\}.$$

(iii) Under both of the assumptions in (i) and (ii),

$$g(\alpha^*) \le -\frac{1}{4} \min\left\{\frac{1}{3p + 1}, \frac{1}{6q + 6}\right\} \sigma^{(q-1)/(2q)}.$$

We remark that in Theorem 4.4.8, we only provide a theoretical guarantee for the decrease of the proximity with a carefully chosen step size. In the practical implementation of IPMs, one might prefer to use various line search techniques to find a more aggressive step size that reduces the proximity much more than our conservative prediction.

4.4.3 Complexity of the Algorithm for CPs

We summarize the complexity of the algorithm in this last subsection. Suppose that the present iterate is in the neighborhood $\mathcal{N}(\tau, \mu)$, that is, $\Psi(v) \le \tau$. Then the algorithm will update the parameter μ by $\mu := (1 - \theta)\mu$. Note that after such an update, the proximity $\Psi(v)$ might increase. As we showed in the LO case (see inequality (3.14)), the proximity after the update still satisfies $\Psi(v) \le \psi_0(\theta, \tau, n)$, where $\psi_0(\theta, \tau, n)$ is given by (3.40). By using Theorem 4.4.8 directly and following a similar procedure to the discussion in the LO chapter, we can get the following results step by step.

Lemma 4.4.9 *Suppose* $\Psi(xs, \mu) \le \tau$ *and* $\tau \ge \nu_1^{-1}$. *Then after an update of the barrier parameter, no more than*

$$\left\lceil \frac{8q\nu_1^{-(q-1)/(2q)}}{\gamma \nu_7(q + 1)}(\psi_0(\theta, \alpha, n))^{(q+1)/(2q)} \right\rceil$$

iterations are needed to recenter.

Note that in some special cases, the bounds on the number of iterations in the above lemma can be simplified. For example, if $\kappa = \mathcal{L} = 0$ and $\gamma = 1$ (or equivalently the CP is linear and monotone) and the proximity $\psi(v)$ used in the algorithm is defined by the function $\psi(t) = \Upsilon_{p,q}(t)$ with $\nu_1 = \nu_2 = 1$, then the bound in Lemma 4.4.9 reduces to

$$\left\lceil \frac{8q}{q+1} \max\{3p+1, 6q+6\}(\psi_0(\theta, \tau, n))^{(q+1)/(2q)} \right\rceil.$$

As a direct consequence of Lemma 4.4.9 and Proposition 1.3.2, the total number of iterations is bounded as follows.

Theorem 4.4.10 *If $\tau \geq \nu_1^{-1}$, the total number of iterations required by the algorithm is not more than*

$$\left\lceil \frac{8q\nu_1^{-(q-1)/(2q)}}{\gamma\nu_7(q+1)} (\psi_0(\theta, \alpha, n))^{(q+1)/(2q)} \right\rceil \left\lceil \frac{1}{\theta} \log \frac{n}{\epsilon} \right\rceil.$$

Again, we can utilize the conclusions in Lemma 4.4.7 to simplify the iteration bound in the above theorem. By way of example, we consider the special case where the underlying CP is linear and monotone and the proximity $\psi(v)$ used in the algorithm is given by the function $\psi(t) = \Upsilon_{p,q}(t)$ with $\nu_1 = \nu_2 = 1$, then the iteration bound in Theorem 4.4.10 becomes

$$\left\lceil \frac{8q}{q+1} \max\{3p+1, 6q+6\}(\psi_0(\theta, \tau, n))^{(q+1)/(2q)} \right\rceil \left\lceil \frac{1}{\theta} \log \frac{n}{\epsilon} \right\rceil.$$

In the specific case of monotone LCPs, the results in this section are only slightly weaker (with a little different constant in the estimation) than the ones in its LO analogue. This is because in the present chapter, we are aiming at establishing the iteration bound for large classes of CPs. Because of the nonlinearity of the underlying mapping, it is very hard to estimate the second derivative of $g(\alpha)$. Therefore, in our analysis we deal with only the first derivative $g'(\alpha)$ (or $g_1'(\alpha)$), which is relatively easier to bound. However, in the case of LCPs, the second derivative $g_1''(\alpha)$ can be estimated similarly by following the approach presented in Chapter 3. In this situation, a slightly sharper estimate of the decrease of the proximity after one step can be obtained and then we can show that the new IPMs for monotone LCPs enjoy the same complexity results as their LO counterparts. However, we are not sure whether such analysis can be extended to nonlinear and non-monotone CPs, while the analysis in this chapter can indeed be applied to the LO case. Finally we remark that from its definition (3.40) one can conclude that $\psi_0(\theta, \tau, n) = \mathcal{O}(n)$ if θ is a constant in $(0,1)$ and $\tau = \mathcal{O}(n)$. In such a situation, if we choose the kernel function $\psi(t) = \Upsilon_{p,q}(t)$ (2.5) with $q = \log n$ and $p \geq 1$ a constant, then from Theorem 4.4.10 we can claim that the algorithm has $\mathcal{O}\left(\sqrt{n}\log n \log(n/\varepsilon)\right)$ iteration bound. This gives the best iteration bound to date of large-update IPMs for $P_*(\kappa)$ CPs.

Chapter 5

Primal-Dual Interior-Point Methods for Semidefinite Optimization Based on Self-Regular Proximities

This chapter considers primal-dual algorithms for solving SDO problems based on self-regular functions defined on the positive definite cone $S_{++}^{n \times n}$. We start with a brief survey of major developments in the area of SDO, after which the notion of self-regular functions is generalized to the case of positive definite cones. Then, self-regular proximities for SDO are introduced as a combination of self-regular functions in $S_{++}^{n \times n}$ joint with the NT scaling. Several fundamental properties of self-regular proximities for SDO are described. The first and second derivatives of a function involving matrix functions are also estimated. We then propose some new search directions for SDO based on self-regular proximities. These new primal-dual IPMs for SDO enjoy the same complexity as their LO counterparts.

5.1 INTRODUCTION TO SDO, DUALITY THEORY AND CENTRAL PATH

The first paper dealing with SDO problems dates back to the early 1960s [11]. For the next many years, the whole topic of SDO stayed silent except for a few isolated results scattered in the literature. The situation changed dramatically around the beginning of the 1990s when SDO started to emerge as one of the fastest developing areas of mathematical programming.

Several reasons can be given why SDO was out of interest for such a long time. One of the main reasons was the lack of robust and efficient algorithms for solving SDO before 1990s. The thrust of SDO research in the 1990s is rooted in two aspects. The first is the importance of the problem. The SDO paradigm not only contains many important class of problems in the field of mathematical programming including LO, QP, SOCO, LCPs, but is also widely used in other fields such as control theory, signal processing, statistics and combinatorial optimization. Interested readers are referred to [55,103,118, 120] for more concrete descriptions. Second, around the end of the 1980s, invigorated by the great achievements of IPMs in coping with problems such as LO, QP and LCPs, the investigation of efficient IPMs for solving SDO reached the top of the agenda of many experts in the field.

According to the records, Nesterov and Nemirovskii [83] were the first to extend IPMs from LO to SDO and build up the polynomial complexity of the algorithm, at least in a theoretical sense. Independently, Alizadeh [2] elegantly applied IPMs to solving SDO problems arising from combinatorics. Since then, a large number of results have been reported concerning two main aspects of SDO research: one is the theoretical analysis and efficient implementation of IPMs; the other is exploration of new applications. As the outcome of much effort in the first direction, the theory of IPMs for SDO has been developed into a mature discipline and quite a number of effective codes [3,26,104,112,116] have been designed in recent years with which most moderate-sized SDO problems can noe be solved successfully, although a general-purpose large-scale SDO solver remains an elusive goal. With regard to the second aspect, many exciting new results about the quality of relaxations of combinatorial and global optimization problems have been reported [30,82]. Even today, new algorithms and applications for SDO are still emerging. The book edited by Wolkowicz, Saigal and Vandenberghe [120] collects review papers on various topics associated with SDO and a large number of references. For more recent achievements, visit the online web site *www.mcs.anl.gov/otc/InteriorPoint* maintained by S. Wright, or the SDP Homepage *http://www.zib.de/helmberg/semidef.html* by Helmberg.

In this chapter we undertake the task of extending the algorithms posed in the previous chapter to the case of SDO. We consider the SDO problem in the following standard form:

$$\text{(SDO)} \quad \min \text{Tr}(CX)$$

$$\text{Tr}(A_i X) = b_i, \quad i = 1, \ldots, m, \ X \succeq 0,$$

and its dual problem

$$\text{(SDD)} \quad b^T y$$

$$\sum_{i=1}^{m} y_i A_i + S = C, S \succeq 0.$$

Here C and A_i $(1 \leq i \leq m)$ are symmetric $n \times n$ matrices, and $b, y \in \Re^m$. Furthermore, "$X \succeq 0$" (or $X \succ 0$) means that X is a symmetric positive semi-definite (or positive definite) matrix. The matrices A_i are assumed to be line-arly independent. SDO reduces to LO in the special case where all the matrices A_i and C are diagonal, which further implies that both X and S can be assumed to be diagonal in this case, because the off-diagonal elements of X and S do not play any role in the objective function or constraints. Many theoretical results about IPMs for LO have been transparently extended to their SDO analogue. For instance, if the primal-dual pair (X, y, S) is feasible for the primal-dual model (SDO) and (SDD), then $b^T y \leq \text{Tr}(CX)$. This is a direct extension of its LO counterpart. However, as pointed out by various authors [55,103,120], for SDO it is impossible to build up the duality theory as strongly as Theorem 1.2.2 for LO. Nevertheless, we have the following relatively weaker duality theory [83,118,103].

Theorem 5.1.1 *We have a pair of optimal solutions X^*, (y^*, S^*) satisfying $\text{Tr}(CX^*) = b^T y^*$ if both of the following conditions hold:*

(i) (SDO) is strictly feasible, that is, there exists an $X \in S^{n \times n}$ such that

$$\text{Tr}(A_i X) = b_i, \quad i = 1, \ldots, m, \quad X \succ 0.$$

(ii) (SDD) is strictly feasible, that is, there exists $S \in S^{n \times n}, y \in \Re^m$ such that

$$\sum_{i=1}^{m} y_i A_i + S = C, \quad S \succ 0.$$

Further, if condition (i) holds and (SDO) is bounded from below, then there is a dual optimal solution (y^*, S^*) satisfying

$$b^T y^* = \inf\{\text{Tr}(CX) : \text{Tr}(A_i X) = b_i, \ i = 1, \ldots, m; \ X \succeq 0\};$$

if condition (ii) holds and (SDO) is bounded from above, then there is a primal optimal solution X^* satisfying

$$\text{Tr}(CX^*) = \sup\left\{b^\mathsf{T}y : \sum_{i=1}^{m} y_i A_i + S = C, \ S \succeq 0\right\}.$$

More details about the duality theory of SDO can be found in the book [120], the thesis by Sturm [103], and the references listed there. For our present purposes, the above theorem is enough. Note that from Theorem 5.1.1 we obtain that, if both (SDO) and (SDD) are strictly feasible, then there exists an optimal solution (X^*, y^*, S^*) of the primal-dual pair (SDO) and (SDD) satisfying

$$\text{Tr}(CX^*) - b^\mathsf{T}y^* = \text{Tr}\left(X^*(S^* + \sum_{i=1}^{m} y_i^* A_i)\right) - b^\mathsf{T}y^*$$

$$= \text{Tr}(X^*S^*) + \sum_{i=1}^{m} y_i^* \text{Tr}(A_i X^*) - b^\mathsf{T}y^* = \text{Tr}(X^*S^*) = 0.$$

Since both X^* and S^* are positive semidefinite, the above relation gives

$$X^*S^* = 0.$$

The concept of the central path has also been extended to SDO. This was first done by Nesterov and Nemirovskii [83] via the extended logarithmic barrier $\log \det(X)$ for the positive semidefinite constraint. Throughout this chapter, since we are concerned with the complexity theory of new algorithms, we assume that both the SDO and its dual SDD are strictly feasible. The central path for SDO is defined by the solution sets $\{X(\mu), y(\mu), S(\mu) : \mu > 0\}$ of the system

$$\begin{cases} \text{Tr}(A_i X) = b_i, & i = 1, \ldots, m, \\ \sum_{i=1}^{m} y_i A_i + S = C, \\ XS = \mu E, & X, S \succeq 0, \end{cases} \tag{5.1}$$

where E denotes the $n \times n$ identity matrix and $\mu > 0$. The following theorem establishing the uniqueness and existence of the central path is well known [83,61,117].

Theorem 5.1.2 *Suppose that both (SDO) and (SDD) are strictly feasible. Then, for every positive μ, there is a unique solution $(X(\mu), y(\mu), S(\mu))$ in $S_{++}^{n \times n} \times \mathfrak{R}^m \times S_{++}^{n \times n}$ to the system (5.1).*

As one of the existing topics pertinent to SDO, the limiting behavior of the central path $(X(\mu), S(\mu))$ has been investigated by many researchers. In particular, we say a primal-dual solution (X^*, S^*) of the underlying SDO problem is strictly complementary if the matrix $X^* + S^* \succ 0$. However, for general SDO problems, there might not exist a strictly complementary solution pair, which is in contrast with the results presented for LO in chapter 3. Instead of the strict complementary solution, the so-called maximally complementary solutions for strictly feasible SDO do exist. Moreover, it has been shown that the central path converges not only to the optimal solution set of (SDO) and (SDD), but also to a maximally complementary solution pair. For this and other properties about the solution set of SDO, one can consult [31,55,103, 120] and the references therein for more details. The basic idea of IPMs is to follow this central path and approximate the optimal set of SDO as μ goes to zero. The strategies, a general procedure and the search direction used in our algorithm for SDO are discussed in the later sections.

5.2 PRELIMINARY RESULTS ON MATRIX FUNCTIONS

Because of the special structure of the underlying problem, in the analysis of IPMs for SDO one usually appeals to some knowledge in matrix theory for help. In this section, we give some preliminary results about matrix functions that are used repeatedly in our later analysis in this chapter. Let us first introduce the definition of a matrix function [10].

Definition 5.2.1 *Suppose the matrix X is diagonalizable with eigen-decomposition*

$$X = Q_X^{-1} \, \mathrm{diag}(\lambda_1(X), \lambda_2(X), ..., \lambda_n(X))Q_X,$$

where Q_X is nonsingular, and let $\psi(t)$ be a function from $\Re$ into itself. The function $\psi(X)$ is defined by

$$\psi(X) = Q_X^{-1} \, \mathrm{diag}(\psi(\lambda_1(X)), \psi(\lambda_2(X)), ..., \psi(\lambda_n(X)))Q_X. \tag{5.2}$$

In particular, if X is symmetric then Q_X can be chosen to be orthogonal, that is, $Q_X^{-1} = Q_X^{\mathrm{T}}$.

Remark 5.2.3 In the rest of this chapter, when we use the function $\psi(\cdot)$ or $h(\cdot)$ and its derivative function $\psi'(\cdot)$ or $h'(\cdot)$ without any specification, it always denotes a matrix function if the argument is a matrix and it means a general function from $\Re$ to $\Re$ if the argument is also in $\Re$. According to (5.2), for any symmetric $X \succeq 0$, we can write $X^{1/2}$ as the square root of X. Similarly we can define the power X for any $X \succ 0$ and $\in \Re$. We also call $\psi(t)$ the kernel function for the matrix function $\psi(X)$.

Now let us first recall Definition 2.1.1 from Chapter 2. By using definition (5.2) of a matrix function we can introduce the notion of self-regular function on the cone of positive definite matrices.

Definition 5.2.3 *A matrix function $\psi(X)$ given by (5.2) is self-regular on $S_{++}^{n \times n}$ if the kernel function $\psi(t)$ is self-regular.*

As we have done in the LO case, for any diagonalizable matrix G with real eigenvalues, we denote by $\Psi(G)$ the trace of the matrix function $\psi(G),$[1] that is,

$$\Psi(G) = \sum_{i=1}^{n} \psi(\lambda_i(G)) = \text{Tr}(\psi(G)), \qquad (5.3)$$

where $\psi(G)$ is defined by (5.2).

We insert here two technical results describing the behavior of a function satisfying condition SR2 (2.2) in the case where the argument of the function is defined by the singular values and eigenvalues of matrices. The lemma is a direct consequence of conclusions (d) of both Theorems 3.3.13 and 3.3.14 in [40, pp. 175–177]. For ease of reference we quote it here without proof. Interested readers can consult [40] for details.

Lemma 5.2.4 *Let $G, H \succ 0$ be given with eigenvalues such that*

$$\lambda_1(G) \geq \lambda_2(G) \geq \cdots \geq \lambda_n(G) > 0, \quad \lambda_1(H) \geq \lambda_2(H) \geq \cdots \geq \lambda_n(H) > 0.$$

Let $\lambda_i(GH), \varrho_i(GH), i \in I$ be the ordered eigenvalues and singular values of the matrix GH, respectively. If $\psi(t)$ is a function satisfying condition SR2, then

$$\sum_{i=1}^{n} \psi(\lambda_i(GH)) \leq \sum_{i=1}^{n} \psi(\varrho_i(GH)) \leq \sum_{i=1}^{n} \psi(\lambda_i(G)\lambda_i(H))$$

$$\leq \frac{1}{2} \sum_{i=1}^{n} (\psi(\lambda_i(G^2)) + \psi(\lambda_i(H^2))).$$

Note that, as we have shown in Lemma 2.1.3, if a function $\psi(t)$ satisfies condition SR2, then so does the function $\psi(t^{\rho})$. By applying Lemma 5.2.4 to the function $\psi(t^{\rho})$, we obtain the following corollary.

Corollary 5.25 *Let $G, H \succ 0$ be given with eigenvalues such that*

$$\lambda_1(G) \geq \lambda_2(G) \geq \cdots \geq \lambda_n(G) > 0, \quad \lambda_1(H) \geq \lambda_2(H) \geq \cdots \geq \lambda_n(H) > 0.$$

Let $\lambda_i(GH), \varrho_i(GH), i \in I$ be the ordered eigenvalues and singular values of the matrix GH, respectively. If $\psi(t)$ is a function satisfying condition SR2, then for any $\rho \in \Re$, the relation

$$\sum_{i=1}^{n} \psi(\lambda_i^{\rho}(GH)) \leq \sum_{i=1}^{n} \psi(\varrho_i^{\rho}(GH)) \leq \sum_{i=1}^{n} \psi(\lambda_i^{\rho}(G)\lambda_i^{\rho}(H))$$

---

[1] G might not be symmetric positive definite.

$$\leq \frac{1}{2} \sum_{i=1}^{n} (\psi(\lambda_i(G^{2\rho})) + \psi(\lambda_i(H^{2\rho})))$$

holds.

Our next proposition presents several appealing characterizations of a self-regular function in $S_{++}^{n \times n}$.

Proposition 5.2.6   *Let the functions $\psi(X) : S_{++}^{n \times n} \to S_{++}^{n \times n}$ and $\Psi(X) : S_{++}^{n \times n} \to \mathfrak{R}$ be defined by (5.2) and (5.3), respectively. If $\psi(X)$ is a self-regular function in $S_{++}^{n \times n}$, then the following statements hold:*

(i) *$\Psi(X)$ is strictly convex with respect to $X \succ 0$ and vanishes at its global minimal point $X = E$, that is, $\Psi(E) = 0, \psi(E) = \psi'(E) = 0_{n \times n}$. Further there exist two positive constants $\nu_1, \nu_2 > 0$ such that*

$$\nu_1(X^{p-1} + X^{-1-q}) \preceq \psi''(X) \preceq \nu_2(X^{p-1} + X^{-1-q}), \quad p, q \geq 1. \quad (5.4)$$

(ii) *For any $X_1, X_2 \succ 0$,*

$$\Psi\left([X_1^{1/2}X_2X_1^{1/2}]^{1/2}\right) \leq \frac{1}{2}(\Psi(X_1) + \Psi(X_2)). \quad (5.5)$$

Proof   To show that the first statement in the proposition is true, we must prove that $\Psi(X)$ is strictly convex for $X \succ 0$; that is for any $X_1, X_2 \succ 0$ and $X_1 \neq X_2$,

$$\Psi\left(\frac{X_1 + X_2}{2}\right) < \frac{1}{2}(\Psi(X_1) + \Psi(X_2)).$$

Since both X_1 and X_2 are positive definite, so is the matrix $(X_1 + X_2)/2$. Let Q be the orthogonal matrix that diagonalizes $(X_1 + X_2)/2$, that is,

$$\Lambda = \text{diag}(\lambda) = \frac{1}{2}Q(X_1 + X_2)Q^{\mathrm{T}} = \frac{1}{2}(Q_1\Lambda_1Q_1^{\mathrm{T}} + Q_2\Lambda_2Q_2^{\mathrm{T}}), \quad (5.6)$$

where both Q_1, Q_2 are orthogonal, $\Lambda_1 = \text{diag}(\lambda^1), \Lambda_2 = \text{diag}(\lambda^2)$ are positive diagonal matrices, and $\lambda, \lambda^1, \lambda^2$ are vectors whose components are the eigenvalues of $(X_1 + X_2)/2, X_1$ and X_2 respectively. Let G be the matrix whose entries are defined by $G_{ij} = [Q_1]_{ij}^2$ and similarly let H have entries $H_{ij} = [Q_2]_{ij}^2$. From (5.6) we readily obtain

$$\lambda = \frac{1}{2}(G\lambda^1 + H\lambda^2).$$

Now, making use of the orthogonality of the matrices Q_1 and Q_2, one can easily see that G and H are two doubly stochastic matrices whose entries satisfy the following relations:

$$\sum_{i=1}^{n} G_{ij} = \sum_{i=1}^{n} H_{ij} = 1, \quad j = 1, 2, ..., n;$$

$$\sum_{j=1}^{n} G_{ij} = \sum_{j=1}^{n} H_{ij} = 1, \quad i = 1, 2, ..., n.$$

Hence

$$\Psi\left(\frac{X_1 + X_2}{2}\right) = \sum_{i=1}^{n} \psi(\lambda_i) = \sum_{i=1}^{n} \psi\left(\frac{G_i\lambda^1 + H_i\lambda^2}{2}\right)$$

$$< \frac{1}{2}\left(\sum_{i=1}^{n} \psi(G_i\lambda^1) + \sum_{i=1}^{n} \psi(H_i\lambda^2)\right),$$

where the inequality follows from the strict convexity of $\psi(t)$ and the assumption that $X_1 \neq X_2$. Using the convexity of $\psi(t)$ again, we have

$$\sum_{i=1}^{n} \psi(G_i\lambda^1) \leq \sum_{i=1}^{n}\sum_{j=1}^{n} G_{ij}\psi(\lambda_j^1) = \sum_{j=1}^{n} \psi(\lambda_1^j) = \Psi(X_1),$$

where the first equality is given by the doubly stochastic nature of the matrix G. Similarly,

$$\sum_{i=1}^{n} \psi(H_i\lambda^2) \leq \Psi(X_2).$$

The above two inequalities yield the desired relation (5.6). From (5.3) and (5.2) one can easily verify all other claims in the first statement.

We next turn to the second statement of the proposition. Using the notation introduced earlier, for any nonsingular matrix $G \in \Re^{n \times n}$ and any $i = 1, ..., n$,

$$\varrho_i(G) = (\lambda_i(G^T G))^{1/2} = (\lambda_i(GG^T))^{1/2}.$$

It follows immediately that for any $X, S \succ 0$,

$$\varrho_i(X^{1/2}S^{1/2}) = (\lambda_i(X^{1/2}SX^{1/2}))^{1/2} = \lambda_i([X^{1/2}SX^{1/2}]^{1/2}).$$

Hence, invoking the definition of $\Psi(X)$, one has

$$\Psi\left([X^{1/2}SX^{1/2}]^{1/2}\right) = \sum_{i=1}^{n} \psi\left(\varrho_i(X^{1/2}S^{1/2})\right).$$

Now recalling Lemma 5.2.4, and replacing the arguments G, H in Lemma 5.2.4 by $X^{1/2}$ and $S^{1/2}$ respectively, we obtain

$$\Psi\left([X^{1/2}SX^{1/2}]^{1/2}\right) = \sum_{i=1}^{n}\psi\left(\varrho_i(X^{1/2}S^{1/2})\right) \le \sum_{i=1}^{n}\psi\left(\varrho_i(X^{1/2})\varrho_i(S^{1/2})\right)$$

$$\le \frac{1}{2}\sum_{i=1}^{n}\left(\psi(\varrho_i^2(X^{1/2})) + \psi(\varrho_i^2(S^{1/2}))\right)$$

$$= \frac{1}{2}\sum_{i=1}^{n}(\psi(\lambda_i(X)) + \psi(\lambda_i(S)))$$

$$= \frac{1}{2}(\Psi(X) + \Psi(S)),$$

where the second inequality follows from condition SR2, and the second equality is implied by the fact that both X and S are symmetric positive definite. $\square$

The two statements (i) and (ii) of Proposition 5.2.6 can be viewed as transparent extensions of conditions SR1 and SR2 introduced in Chapter 2, respectively.

We continue our discussion about matrix functions as follows. First note that, as we mentioned earlier in Chapter 1, to analyze IPMs, we usually need to study the behavior of various potential functions or proximity measures. In the case of SDO, such proximity measures are always defined as a function with matrix variables. In general it is very difficult to compute the gradient of a matrix function, not to mention its Hessian. However, for our specific purpose, we do not really need the gradient of the matrix function. Instead, we compute the derivatives of a function involving matrix functions. To continue our analysis, we need another concept relevant to matrix functions in matrix theory. This is the notion of matrices of functions (or matrix-valued functions) [40].

Definition 5.2.7   *A matrix $X(t)$ is said to be a matrix of functions (or a matrix-valued function) if each entry of $X(t)$ is a function of t, that is, $X(t) = [X_{ij}(t)]$.*

The usual concepts of continuity, differentiability and integrability have been extended to matrix-valued functions of a scalar by interpreting them entry-wise. Thus we have

$$\frac{d}{dt}X(t) = \left[\frac{d}{dt}X_{ij}(t)\right].$$

For simplicity, in the rest of this section we denote by $X'(t)$ the gradient of the matrix-valued function $X(t)$. The following relations between matrix-valued functions and their derivatives [40, pp. 490–491] are well known in the theory of matrices. These relations are very helpful in the proof of our next lemma. Suppose that the matrix-valued functions $G(t), H(t)$ are differentiable and nonsingular at the point t. Then we have

$$\frac{d}{dt}\mathrm{Tr}\,(G(t)) = \mathrm{Tr}\left(\frac{d}{dt}G(t)\right) = \mathrm{Tr}\,(G'(t)), \qquad (5.7)$$

$$\frac{d}{dt}(G(t)H(t)) = \left[\frac{d}{dt}G(t)\right]H(t) + G(t)\left[\frac{d}{dt}H(t)\right] = G'(t)H(t) + G(t)H'(t).$$
$$(5.8)$$

For any function $\psi(t)$, let us denote by $\Delta\psi$ the divided difference of $\psi(t)$ such that

$$\Delta\psi(t_1, t_2) = \frac{\psi(t_1) - \psi(t_2)}{t_1 - t_2} \quad \forall t_1 \neq t_2 \in \Re,$$

and in the particular case $t = t_1 = t_2$, we define $\Delta\psi(t, t) = \psi'(t)$. The following result provides some ways to evaluate the first derivative of a function involving matrices of functions and to bound its second derivative.

Lemma 5.2.8 *Suppose that $H(t)$ is a matrix of functions such that the matrix $H(t)$ is symmetric and diagonalizable with eigenvalues $\lambda_1(t) \geq \lambda_2(t) \geq \cdots \geq \lambda_n(t)$. If $H(t)$ is twice differentiable with respect to t for all $t \in (l_t, u_t)$ and $\psi(t)$ is also a twice continuously differentiable function in a suitable domain that contains all the eigenvalues of $H(t)$, then*

$$\frac{d}{dt}\mathrm{Tr}(\psi(H(t))) = \mathrm{Tr}(\psi'(H(t))H'(t)), \quad t \in (l_t, u_t),$$

and

$$\frac{d^2}{dt^2}\mathrm{Tr}(\psi(H(t))) \leq \varpi\|H'(t)\|^2 + \mathrm{Tr}(\psi'(H(t))H''(t)), \qquad (5.9)$$

where

$$\varpi = \max\{|\Delta\psi'(\lambda_j(t), \lambda_k(t))|, \; j, k = 1, 2, ..., n\} \qquad (5.10)$$

is a number depending on $H(t)$ and $\psi(t)$ with

$$\Delta\psi'(t_1, t_2) = \frac{\psi'(t_1) - \psi'(t_2)}{t_1 - t_2} \quad \forall t_1, t_2 \in [l_t, u_t].$$

Proof The proof is essentially an immediate consequence of [40, (6.6.27), pp. 531]. For self-completeness we write it out here. Since $H(t)$ is positive definite, it is diagonalizable. Hence we can assume without loss of generality that $H(t) = Q_H(t)^{\mathrm{T}}\Lambda_H(t)Q_H(t)$, where $\Lambda_H(t)$ is a diagonal matrix whose elements are the eigenvalues $\{\lambda_1(t), ..., \lambda_n(t)\}$ of $H(t)$, and $Q_H(t)$ is orthogonal. For notation simplicity, in the proof of this lemma we omit the parameter t and use the notations Q_H, H, H', λ_j for $Q_H(t)$, $H(t)$, $H'(t)$ and $\lambda_j(t)$, respectively. Further, let us denote by D_j the diagonal matrix that has a 1 in its (j,j) position and all other entries of D_j are zero. It follows from [40, (6.6.27), pp. 531] that[2]

$$\frac{d}{dt}\psi(H(t)) = Q_H^T\left(\sum_{j,k=1}^{n} \Delta\psi(\lambda_j, \lambda_k)D_j[Q_H H' Q_H^T]D_k\right)Q_H. \quad (5.11)$$

Recalling the definition of the trace of a matrix, we know that for any matrices G, H, there holds $\mathrm{Tr}(GH) = \mathrm{Tr}(HG)$. This simple fact, combined with (5.11) and (5.7), gives

$$\mathrm{Tr}\left(\frac{d}{dt}\psi(H(t))\right) = \mathrm{Tr}\left(\sum_{j,k=1}^{n} \Delta\psi(\lambda_j, \lambda_k)D_j\left[Q_H H' Q_H^T\right]D_k\right).$$

Now by the choice of D_j, we can claim that for any symmetric matrix G, there holds $\mathrm{Tr}(D_j G D_k) = 0$ if $\lambda_j(t) \neq \lambda_k(t)$. Thus it readily follows that

$$\mathrm{Tr}\left(\frac{d}{dt}\psi(H(t))\right) = \mathrm{Tr}\left(\sum_{j,k=1}^{n} \Delta\psi(\lambda_j, \lambda_j)D_j\left[Q_H H' Q_H^T\right]D_k\right)$$

$$= \mathrm{Tr}\left(\sum_{j=1}^{n} \Delta\psi(\lambda_j, \lambda_j)D_j\left[Q_H H' Q_H^T\right]D_j\right)$$

$$= \mathrm{Tr}\left(\sum_{j=1}^{n} \psi'(\lambda_j)\left[Q_H^T D_j^2 Q_H\right]H'\right),$$

where the last equality is given by the definition of $\Delta\psi(\cdot, \cdot)$. The above equality implies

$$\mathrm{Tr}\left(\frac{d}{dt}\psi(H(t))\right) = \mathrm{Tr}\left(\sum_{j=1}^{n} Q_H^T(t)[D_j\psi'(\lambda_j)D_j]Q_H H'\right)$$

---

[2] In [40], the authors used the distinct eigenvalues of $H(t)$ to define the block matrix D_j whose (i,i)th entry is 1 if $\lambda_i = \lambda_j$ while the constant n in (5.11) was replaced by the number of distinct eigenvalues of $H(t)$. However, it can be easily verified that our expression (5.11) is equivalent to that given by (6.6.27) in [40, pp. 531]. We introduce (5.11) to be consistent with our definition about a matrix's eigenvalues in the introduction and to avoid too much new notation.

$$= \mathrm{Tr}(Q_H^{\mathrm{T}}\mathrm{diag}(\psi'(\lambda_1), \psi'(\lambda_2), ..., \psi'(\lambda_n))Q_H H')$$

$$= \mathrm{Tr}(\psi'(H)H').$$

This completes the proof of the first conclusion in the lemma.

We now prove the second statement of the lemma. First note that by applying the chain rule (5.8) one easily gets

$$\frac{d^2}{dt^2}\mathrm{Tr}(\psi(H(t))) = \frac{d}{dt}\mathrm{Tr}(\psi'(H(t))H'(t))$$

$$= \mathrm{Tr}(\psi'(H(t))H''(t)) + \mathrm{Tr}\left([\frac{d}{dt}\psi'(H(t))]H'(t)\right).$$

Hence it remains to show that

$$\mathrm{Tr}\left([\frac{d}{dt}\psi'(H(t))]H'(t)\right) \leq \varpi\|H'(t)\|^2.$$

Replacing the function $\psi(t)$ by $\psi'(t)$ in (5.11) we obtain

$$\mathrm{Tr}\left(\left[\frac{d}{dt}\psi'(H(t))\right]H'(t)\right) = \mathrm{Tr}\left(Q_H^{\mathrm{T}}\left(\sum_{j,k=1}^{n}\Delta\psi'(\lambda_j, \lambda_k)D_j\left[Q_H H'Q_H^{\mathrm{T}}\right]D_k\right)Q_H H'\right)$$

$$= \mathrm{Tr}\left(\left(\sum_{j,k=1}^{n}\Delta\psi'(\lambda_j, \lambda_k)D_j[Q_H H'Q_H^{\mathrm{T}}]\right)D_k[Q_H H'Q_H^{\mathrm{T}}]\right).$$

Again, by the choice of D_j, D_k, we can easily verify that for any symmetric matrix G, we have $\mathrm{Tr}(D_j G D_k G) = G_{j,k}^2$. Hence it immediately follows that

$$\mathrm{Tr}\left(\left[\frac{d}{dt}\psi'(H(t))\right]H'(t)\right) = \mathrm{Tr}\left(\sum_{j,k=1}^{n}\Delta\psi'(\lambda_j, \lambda_k)\left[Q_H H'Q_H^{\mathrm{T}}\right]_{j,k}^2\right)$$

$$\leq \varpi\sum_{j,k=1}^{n}\left[Q_H H'(t)Q_H^{\mathrm{T}}\right]_{j,k}^2$$

$$= \varpi\|H'(t)\|^2,$$

where the inequality is implied by the choice of ϖ. $\square$

5.3 NEW SEARCH DIRECTIONS FOR SDO

5.3.1 Scaling Schemes for SDO

Most IPMs for SDO follow the same pattern as their counterparts for LO and use certain strategies to trace the central path appropriately. However, there is a big difference between the search directions used in the LO and SDO algorithms. In the sequel we explain how this distinction arises. Suppose the point (X, y, S) is strictly feasible. Newton's method amounts to linearizing system (5.1), thus yielding the following equation:

$$\text{Tr}(A_i \Delta X) = 0, \qquad i = 1, \dots, m, \tag{5.12}$$

$$\sum_{i=1}^{m} \Delta y_i A_i + \Delta S = 0 \tag{5.13}$$

$$X \Delta S + \Delta X S = \mu E - XS. \tag{5.14}$$

A crucial observation for SDO is that the above Newton system might not have a symmetric solution ΔX. To make up for this point, many researchers have proposed different ways of symmetrizing the third equation in the Newton system so that the new system has a unique symmetric solution. In [109], Todd gave a comprehensive survey about various search directions for SDO and listed a total of more than 20 search directions. Among others, the following three directions are the most popular:

(1) The Alizadeh, Haeberly, Overton (AHO) direction introduced in [4].

(2) The search direction independently proposed by Helmberg, Rendl, Vanderbei and Wolkowicz [38], and Kojima, Shindoh and Hara [61], and later rediscovered by Monteiro [74], which we refer to as the HKM direction.

(3) The Nesterov and Todd (NT) direction [84,85].

In [132], by using a general nonsingular matrix P, Zhang suggested the symmetrization scheme

$$H_P(XS) = \frac{1}{2}\left(PXSP^{-1} + P^{-T}SXP^{T}\right)$$

to get a new symmetric linearized Newton equation,

$$\begin{cases} \text{Tr}(A_i \Delta X) = 0, \ i = 1, \dots, m; \ \Delta S = -\sum_{i=1}^{m} \Delta y_i A_i; \\ P(XS + X\Delta S + \Delta XS)P^{-1} + P^{-T}(SX + S\Delta X + \Delta SX)P^{T} = 2\mu E. \end{cases} \tag{5.16}$$

This scheme was an extension of the symmetrizing trick first introduced by

Monteiro [74], where only some specific cases of P were considered. The class of search directions defined by the solutions of system (5.16) are now referred to as the Monteiro–Zhang (MZ) family. Various authors have studied this system, and certain sufficient conditions have been given under which the linearized Newton system has a unique solution [109,110]. Further, it is known that most efficient search directions for SDPs are solutions of the above system with different P. For instance, the simple argument $P = E$ leads to the (AHO) [4] direction; the choice $P = S^{1/2}$ ($P = X^{-1/2}$) gives the primal (dual) HKM direction [38,61,74], and the choice

$$P_{NT} = W^{-1/2}, \quad W = X^{1/2}(X^{1/2}SX^{1/2})^{-1/2}X^{1/2} = S^{-1/2}(S^{1/2}XS^{1/2})^{1/2}S^{-1/2} \quad (5.17)$$

yields the NT direction [84,85]. There are, of course, some other classes of search directions for solving SDO. Since these other search directions are not so relevant to our discussion, for simplicity we omit them here. Readers interested in those directions may consult the survey paper [109] or the handbook [120].

5.3.2 Intermezzo: A Variational Principle for Scaling

With regard to the choice of scaling scheme, it is clearly desirable that the resulting Newton system enjoys certain appealing features. In [109], Todd posed quite a number of attractive features for the scaled Newton system defining the search directions for SDO. For instance, the system should be able to predict the duality gap,[3] be scale invariant and primal-dual symmetric, etc. However, none of the known search directions satisfies all these properties. On the other hand, even in the simple LO case, one can readily see that some of the new search directions introduced in Chapter 3 do not predict the duality gap exactly.

In this section, among others we consider the symmetrization scheme from which the NT direction [84,105,110] is derived. One important reason for this is that the NT scaling technique transfers the primal variable X and the dual S into the same space: the so-called V-space. Let $D = P_{NT}^{-1} = W^{1/2}$. The matrix D can be used to rescale X and S to the same matrix V [55,105] defined by

$$V := \frac{1}{\sqrt{\mu}} D^{-1}XD^{-1} = \frac{1}{\sqrt{\mu}} DSD. \quad (5.18)$$

Since the matrix W is symmetric and positive definite, so are the matrices D and V. The notation V provides us a very simple way to express the centrality condition, which can be now stated as $V = E$. Moreover, the scaled trace $\mu \text{Tr}(V^2)$ represents the duality gap.

---

[3] The search direction $(\Delta X, \Delta S)$ is said to predict the duality gap if $\text{Tr}(X\Delta S + S\Delta X) = n\mu - \text{Tr}(XS)$.

Another reason for our choice comes from a variational principle, which we present in the sequel. Such a principle is also closely related to self-regular functions in the positive definite cone. For any nonsingular P, let us define

$$\tilde{X} = PXP^{\mathrm{T}}, \quad \tilde{S} = P^{-T}SP^{-1}, \quad \tilde{A}_i = P^{-T}A_iP^{-1}, \quad i = 1, \ldots, m.$$

We consider the primal SDO problem in the scaled space

$$\text{(Scaled SDO)} \quad \min \operatorname{Tr}(\tilde{C}\tilde{X})$$

$$\operatorname{Tr}(\tilde{A}_i\tilde{X}) = b_i, \ i = 1, \ldots, m, \ \tilde{X} \succeq 0,$$

and its dual problem

$$\text{(Scaled SDD)} \quad \max b^{\mathrm{T}}y$$

$$\sum_{i=1}^m y_i\tilde{A}_i + \tilde{S} = \tilde{C}, \quad \tilde{S} \succeq 0.$$

Suppose we use some kind of barrier method to solve both the scaled primal and dual problems. That is, at the present iterate, we want to minimize a specific potential function $\operatorname{Tr}(\tilde{C}\tilde{X}) + \Psi(\tilde{X})$ and maximize $b^{\mathrm{T}}y - \Psi(\tilde{S})$, where $\Psi(\cdot)$ is a barrier function with matrix variable. The scaling matrix P at which the summation $\operatorname{Tr}(\tilde{C}\tilde{X}) - b^{\mathrm{T}}y + \Psi(\tilde{X}) + \Psi(\tilde{S})$ has a global minimal value is particularly attractive for our purpose. Our next result shows that such an optimal scaling is exactly the NT scaling if the corresponding kernel function $\psi(t)$ of the barrier function $\Psi(\cdot)$ satisfies condition SR2 strictly.

Proposition 5.3.1 *Suppose the matrix function $\Psi(\cdot)$ is defined by (5.2) and (5.3). If the function $\psi(t)$ satisfies condition SR2 strictly, that is, $\psi'(t) + t\psi''(t) > 0$ for any $t > 0$, then the function $\operatorname{Tr}(\tilde{C}\tilde{X}) - b^{\mathrm{T}}y + \Psi(\tilde{X}) + \Psi(\tilde{S})$ attains its global minimal value at the point P such that $\tilde{X} = \tilde{S}$.*

Proof Since the trace $\operatorname{Tr}(\tilde{C}\tilde{X})$ is invariant for any nonsingular matrix P, it suffices to show that $\Psi(\tilde{X}) + \Psi(\tilde{S})$ has a global minimizer at the point P^* such that

$$\tilde{X}^* = \tilde{S}^* = P^*XP^{*\mathrm{T}} = P^{*-\mathrm{T}}SP^{*-1}.$$

First we note that, for any $X, S \succ 0$, the NT scaling matrix $P^* = P_{NT}$ guarantees $\tilde{X}^* = \tilde{S}^*$ (see [85,105]), which gives the existence of such a matrix P^*. Thus we need only prove that for this choice of P^*, the value of the function $\Psi(\tilde{X}) + \Psi(\tilde{S})$ is optimal. Because $\tilde{X}^* = \tilde{S}^*$, we obtain

$$\frac{1}{2}\left(\Psi(\tilde{X}^*) + \Psi(\tilde{S}^*)\right) = \Psi(\tilde{X}^*) = \sum_{i=1}^n \lambda_i[\psi(\tilde{X}^*)] = \sum_{i=1}^n \lambda_i^{1/2}[\psi(\tilde{X}^*\tilde{S}^*)]$$

$$= \sum_{i=1}^{n} \lambda_i^{1/2} [\psi(\tilde{X}\tilde{S})] = \sum_{i=1}^{n} \lambda_i \left(\psi([\tilde{S}^{1/2}\tilde{X}\tilde{S}^{1/2}]^{1/2}) \right)$$

$$= \sum_{i=1}^{n} \psi(\varrho_i(\tilde{X}^{1/2}\tilde{S}^{1/2})) \leq \sum_{i=1}^{n} \psi(\varrho_i(\tilde{X}^{1/2})\varrho_i(\tilde{S}^{1/2}))$$

$$< \frac{1}{2} \left(\sum_{i=1}^{n} \psi(\varrho_i(\tilde{X})) + \sum_{i=1}^{n} \psi(\varrho_i(\tilde{S})) \right),$$

where the first inequality follows from Lemma 5.2.4, and the last inequality is implied by the assumption that $\psi(t)$ satisfies SR2 strictly. This completes the proof of the proposition. $\qquad\qquad\square$

We note that if the function $\psi(t)$ is chosen as the classical logarithmic barrier $-\log t$ (which satisfies $\psi'(t) + t\psi''(t) = 0$) or the logarithmic barrier $-\log X$ for the positive definite cone, then $\Psi(\tilde{X}) + \Psi(\tilde{S}) = \Psi(X) + \Psi(S)$ for any nonsingular P. Hence, in some sense, the logarithmic barrier is not a very good choice because it cannot help us to distinguish one scaling scheme from many others. It should also be mentioned that for the primal-dual feasible pair (X, y, S) and any nonsingular P, there holds trivially

$$\mathrm{Tr}(\tilde{C}\tilde{X}) - b^T y = \mathrm{Tr}(\tilde{X}\tilde{S}) = \mathrm{Tr}(XS).$$

5.3.3 New Proximities and Search Directions for SDO

Let us go back to NT scaling (page 112). Let us define

$$\bar{A}_i := \frac{1}{\sqrt{\mu}} DA_i D, \quad i = 1, \dots, m;$$

$$D_X := \frac{1}{\sqrt{\mu}} D^{-1}\Delta X D^{-1}, \quad D_S := \frac{1}{\sqrt{\mu}} D\Delta S D. \tag{5.19}$$

Using the above definitions and (5.18), we can write the condition $XS = \mu E$ as $V = E$. Correspondingly the NT direction is defined as the solution of the system

$$\mathrm{Tr}(\bar{A}_i D_X) = 0, \quad i = 1, \dots, m,$$

$$\sum_{i=1}^{m} \Delta y_i \bar{A}_i + D_S = 0, \tag{5.20}$$

$$D_X + D_S = V^{-1} - V.$$

The proximity measure we use here for SDO is defined by

$$\Phi(X, S, \mu) := \Psi(V) = \mathrm{Tr}(\psi(V)). \tag{5.21}$$

In the rest of this chapter, we discuss only the case that the proximity $\psi(V)$ is *self-regular*.

Analogous to the case of LO, the new search direction we suggest for SDO is a slight modification of the NT direction. It is defined by the solution $(D_X, D_S, \Delta y)$ of the following system:

$$\mathrm{Tr}(\overline{A}_i D_X) = 0, \quad i = 1, \ldots, m,$$

$$\sum_{i=1}^{m} \Delta y_i \overline{A}_i + D_S = 0, \tag{5.22}$$

$$D_X + D_S = -\psi'(V).$$

Having D_X and D_S, we can calculate ΔX and ΔS by (5.19). From the orthogonality of ΔX and ΔS, it is trivial to see that

$$\mathrm{Tr}(D_X D_S) = \mathrm{Tr}(D_S D_X) = 0. \tag{5.23}$$

For notational brevity, we also define

$$\sigma^2 = \mathrm{Tr}(\psi'(V)^2) = \|\psi'(V)\|^2 = \|D_X + D_S\|^2 = \|D_X\|^2 + \|D_S\|^2, \tag{5.24}$$

where the last equality is given by (5.23).

Replacing v_i, $v_{\max}$ and $v_{\min}$ by $\lambda_i(V)$, $\lambda_{\max}(V)$ and $\lambda_{\min}(V)$, respectively if necessary, and by a chain of arguments similar to the proof of Proposition 3.1.5, one can prove the following result. These conclusions can also be viewed as common features of general self-regular functions in the cone of positive definite matrices.

Proposition 5.3.2 *Let the proximity* $\Psi(V)$ *be defined by (5.21) and* σ *by (5.24). If the kernel function* $\psi(t)$ *satisfies condition SR1, then*

$$\Psi(V) \leq \frac{\sigma^2}{\nu_1}, \tag{5.25}$$

$$\lambda_{\min}(V) \geq \left(1 + \frac{q\sigma}{\nu_1}\right)^{-1/q}, \tag{5.26}$$

and

$$\lambda_{\max}(V) \leq \left(1 + \frac{p\sigma}{\nu_1}\right)^{1/p}. \tag{5.27}$$

If $\lambda_{\max}(V) > 1$ *and* $\lambda_{\min}(V) < 1$, *then*

$$\sigma \geq \nu_1 \left(\frac{(\lambda_{\max}(V)^p - 1)^2}{p^2} + \frac{(\lambda_{\min}(V)^{-q} - 1)^2}{q^2} \right)^{1/2}. \qquad (5.28)$$

In the sequel we consider the relations between the duality gap and the proximity. Note that, replacing v_i by $\lambda_i(V)$, in a similar vein to the proof of Lemma 2.2.3 we can easily deduce

$$\frac{\Psi(V)}{\nu_1} \geq \frac{1}{2} \sum_{i=1}^n (\lambda_i(V) - 1)^2 = \frac{1}{2} \|V\|^2 - \sum_{i=1}^n \lambda_i(V) + \frac{n}{2} \geq \frac{1}{2} \|V\|^2 - \sqrt{n} \|V\| + \frac{n}{2},$$

which further implies

$$\|V\| \leq \sqrt{n} + \sqrt{\frac{2\Psi(V)}{\nu_1}}.$$

It follows immediately that

$$\mathrm{Tr}(XS) = \mu \|V\|^2 \leq n\mu + 2\mu \sqrt{\frac{n\Psi(V)}{\nu_1}} + \frac{\Psi(V)}{\nu_1} \mu. \qquad (5.29)$$

This implies that the proximity can be used as a potential function for minimizing the duality gap.

We conclude the present section with the following result concerning the proximities $\Psi(V)$, $\Psi(X(\mu)^{-1}X)$ and $\Psi(S(\mu)^{-1}S)$, where $X(\mu)$ and $S(\mu)$ are the targeted centers in the primal and dual space, respectively. This result is a direct consequence of Corollary 5.2.5.

Lemma 5.3.3 *Suppose $X, S \succ 0$ and V is the scaled matrix defined by (5.18). If the kernel function $\psi(t)$ defining (5.3) satisfies condition SR2, then*

$$\Psi(V) \leq \frac{1}{2} \left(\Psi(X(\mu)^{-1}X) + \Psi(S(\mu)^{-1}S) \right).$$

Proof By (5.18) we have

$$V = \frac{1}{\sqrt{\mu}} \left(D^{-1} XSD \right)^{1/2}.$$

The matrix $D^{-1}XSD$ is symmetric positive definite and has the same eigenvalues as the matrix XS. Recalling the fact that $X(\mu)S(\mu) = \mu E$, we can claim immediately that the eigenvalues of the matrix V^2 are precisely the same as those of the matrix $X(\mu)^{-1/2}XX(\mu)^{-1/2}S(\mu)^{-1/2}SS(\mu)^{-1/2}$. Hence, by using Corollary 5.2.5 and Definition 5.31, we obtain

$$\Psi(V) = \sum_{i=1}^{n} \psi\Big(\lambda_i^{1/2}\Big(X(\mu)^{-1/2}XX(\mu)^{-1/2}S(\mu)^{-1/2}SS(\mu)^{1/2}\Big)\Big)$$

$$\leq 1/2\Big(\Psi\Big(X(\mu)^{-1/2}XX(\mu)^{-1/2}\Big) + \Psi\Big(S(\mu)^{-1/2}SS(\mu)^{-1/2}\Big)\Big)$$

as required.   □

5.4 NEW POLYNOMIAL PRIMAL-DUAL IPMS FOR SDO

5.4.1 The Algorithm

In the present section we describe the new primal-dual algorithm for solving SDO. Bearing in mind that the centrality condition for SDO can be written as $V = E$, naturally we also use V to define the neighborhood of the central path. Let us denote by $\mathcal{F}_{SDO}$ the feasible set of the primal-dual SDO, that is,

$$\mathcal{F}_{SDO} = \left\{(X,S) : \text{Tr}(A_i X) = b_i, \ i = 1,...,m; \ \sum_{i=1}^{m} y_i A_i + S = C, \ X, S \succeq 0\right\}.$$

The neighborhood of the central path we use can be defined as follows:

$$\mathcal{N}(\tau, \mu) = \{(X,S) : (X,S) \in \mathcal{F}_{SDO}, X, S \succ 0, \Phi(X,S,\mu) = \Psi(V) \leq \tau\}.$$

Starting from a well-centralized point (which can always be obtained via the so-called self-dual embedding model, as described in [55,120] or Chapter 7), we solve system (5.22) and get a search direction. Then the iterate is updated by certain rules. If the present point is outside some kind of neighborhood, we repeat the inner iterations until the new iterate enters the neighborhood again. Otherwise we continue by performing an outer iteration and reduce μ by a fixed factor. The process is repeated until μ is sufficiently small and an approximate solution of the considered SDO problem is reported.

We stipulate that the algorithm terminates with a point satisfying $n\mu < \varepsilon$. By using (5.29), we obtain that

$$\text{Tr}(XS) \leq n\mu + 2\mu\sqrt{\frac{n\tau}{\nu_1}} + \mu\frac{\tau}{\nu_1}.$$

Hence if $\tau = \mathcal{O}(n)$, which means that the algorithm indeed works in a large neighborhood of the central path, then the algorithm finally reports a feasible solution such that $\text{Tr}(XS) = \mathcal{O}(\varepsilon)$.

The procedure of the algorithm is recorded as follows.

Primal-Dual Algorithm for SDO

Inputs
  A proximity parameter $\tau \geq \nu_1^{-1}$;
  an accuracy parameter $\varepsilon > 0$;
  a fixed barrier update parameter $\theta \in (0, 1)$;
  (X^0, S^0) and $\mu^0 = 1$ such that $\Phi(X^0, S^0, \mu^0) \leq \tau$.
begin
  $X := X^0$; $S := S^0$; $\mu := \mu^0$;
  while $n\mu \geq \varepsilon$ **do**
  begin
    $\mu := (1 - \theta)\mu$;
    while $\Phi(X, S, \mu) \geq \tau$ **do**
    begin
      Solve system (5.22) for $\Delta X, \Delta y, \Delta S$,
      Determine a step size α;
      $X := X + \alpha\Delta X$;
      $y := y + \alpha\Delta y$;
      $S := S + \alpha\Delta S$;
    end
  end
end

5.4.2 Complexity of the Algorithm

Now we can start to estimate the complexity of the algorithm. First observe that, as we have seen in the LO case, the key step in evaluating the complexity of the algorithm is to calculate the decrease of the proximity after one feasible step.

Let us denote by V_+ the scaled matrix defined by (5.17) and (5.18), where the matrices X and S are replaced by $X_+ = X + \alpha\Delta X, S_+ = S + \alpha\Delta S$, respectively. It is trivial to verify that V_+^2 is unitarily similar to the matrix $X_+^{1/2} S_+ X_+^{1/2}$ and thus to the matrix $(V + \alpha D_X)^{1/2}(V + \alpha D_S)(V + \alpha D_X)^{1/2}$. This further implies that the eigenvalues of the matrix V_+ are precisely the same as those of

$$\tilde{V}_+ := \left((V + \alpha D_X)^{1/2}(V + \alpha D_S)(V + \alpha D_X)^{1/2}\right)^{1/2}. \qquad (5.30)$$

Since the proximity after one step is defined by $\Psi(V_+)$, from (5.21), it follows immediately that

$$\Psi(V^+) = \Psi(\tilde{V}_+). \qquad (5.31)$$

Again we define the gap between the proximity before and after one step as a function of the step size; that is,

$$g(\alpha) := \Psi(V^+) - \Psi(V) = \Psi(\tilde{V}^+) - \Psi(V). \tag{5.32}$$

Our goal is to estimate the decreasing value of $g(\alpha)$ for a feasible step size α.
  Let

$$\overline{D}_x = V^{-1/2} D_X V^{-1/2}, \overline{D}_s = V^{-1/2} D_S V^{-1/2}. \tag{5.33}$$

Since

$$V + \alpha D_X = V^{1/2}(E + \alpha \overline{D}_x) V^{1/2}, \quad V + \alpha D_S = V^{1/2}(E + \alpha \overline{D}_s) V^{1/2},$$

the matrices $V + \alpha D_X$ and $V + \alpha D_S$ are positive definite if and only if both the
matrices $E + \alpha \overline{D}_x$ and $E + \alpha \overline{D}_s$ are positive definite. As a consequence, the
maximal feasible step size is dependent on the eigenvalues of matrices $\overline{D}_x$ and
$\overline{D}_s$. Our next result gives some estimates of the norms of the matrices $\overline{D}_x$ and
$\overline{D}_s$, and the maximal feasible step size $\alpha_{\max}$.

Lemma 5.4.1 *Let the matrices $\overline{D}_x$ and $\overline{D}_s$ be defined by (5.33) and $\alpha_{\max}$ be
the maximal feasible step size. Then*

$$\|\overline{D}_x\|^2 + \|\overline{D}_s\|^2 \le \overline{\alpha}^{-2}$$

and

$$\alpha_{\max} \ge \overline{\alpha},$$

where

$$\overline{\alpha} = \sigma^{-1}\left(1 + \frac{q\sigma}{\nu_1}\right)^{-1/q}.$$

Proof First we quote an inequality about the singular values and eigenvalues
of matrices from Section 3.3.20 of [40, pp. 178]. If the matrices G and H are
symmetric, then for any $i \in I$,

$$\varrho_i(GH) \le \min(\varrho_i(G)\varrho_1(H), \varrho_1(G)\varrho_i(H))$$

$$= \min(|\lambda_1(G)\lambda_i(H)|, |\lambda_1(H)\lambda_i(G)|). \tag{5.34}$$

It follows immediately that

$$|\lambda_i(\overline{D}_x)| \le \frac{1}{\lambda_{\min}(V)^{1/2}} \varrho_i(D_X V^{-1/2}) \le \frac{1}{\lambda_{\min}(V)} |\lambda_i(D_X)|, \quad i \in I$$

and

$$|\lambda_i(\overline{D}_s)| \le \frac{1}{\lambda_{\min}(V)^{1/2}} \varrho_i(D_S V^{-1/2}) \le \frac{1}{\lambda_{\min}(V)} |\lambda_i(D_S)|, \quad i \in I.$$

These two relations mean

$$\|\overline{D}_x\|^2 + \|\overline{D}_s\|^2 = \sum_{i=1}^{n} \left(\lambda_i^2(\overline{D}_x) + \lambda_i^2(\overline{D}_s)\right) \leq \frac{1}{\lambda_{\min}(V)^2} \sum_{i=1}^{n} \left(\lambda_i^2(D_X) + \lambda_i^2(D_X)\right)$$

$$= \frac{1}{\lambda_{\min}(V)^2} \left(\|D_X\|^2 + \|D_S\|^2\right) = \frac{1}{\lambda_{\min}(V)^2} \left(\|D_X + D_S\|^2\right)$$

$$\leq \sigma^2 \left(1 + \frac{q\sigma}{\nu_1}\right)^{2/q},$$

where the third equality is given by (5.23), and the last inequality follows from the definition of σ and (5.26). The second statement of the lemma is an immediate consequence of the first result of the lemma. $\qquad \square$

Now we evaluate the function $g(\alpha)$ defined by (5.32). By using the second statement of Proposition 5.2.6, one gets

$$g(\alpha) = \Psi(V^+) - \Psi(V) \leq \frac{1}{2}\left(\Psi(V + \alpha D_X) + \Psi(V + \alpha D_S)\right) - \Psi(V) =: g_1(\alpha).$$

Hence it suffices for us to estimate the decrease of the value of the function $g_1(\alpha)$ after one step. The main difficulty in estimating $g_1(\alpha)$ is to evaluate its first and second derivatives. From Lemma 5.2.8 we get

$$g_1'(\alpha) = \frac{1}{2}\text{Tr}\left(\psi'(V + \alpha D_X)D_X + \psi'(V + \alpha D_S)D_S\right) \qquad (5.35)$$

and

$$g_1''(\alpha) = \frac{1}{2}\frac{d^2}{d\alpha^2}\text{Tr}\left(\psi(V + \alpha D_X) + \psi(V + \alpha D_S)\right). \qquad (5.36)$$

Now we are ready to state one of our main results in this section.

Lemma 5.4.2 *If the step size $\alpha \leq \overline{\alpha}$, then*

$$g_1''(\alpha) \leq \frac{\nu_2\sigma^2}{2}\left((\lambda_{\max}(V) + \alpha\sigma)^{p-1} + (\lambda_{\min}(V) - \alpha\sigma)^{-q-1}\right).$$

Proof First notice that by Lemma 5.4.1 the step size used in the lemma is strictly feasible. From Lemma 5.2.8 we conclude that

$$g_1''(\alpha) \leq \frac{1}{2}\left(\varpi_1\|D_X\|^2 + \varpi_2\|D_S\|^2\right),$$

where

$$\varpi_1 = \max\{|\Delta\psi'(\lambda_j(V + \alpha D_X), \lambda_k(V + \alpha D_X))| : j, k = 1, 2, ..., n\},$$

$$\varpi_2 = \max\{|\Delta\psi'(\lambda_j(V + \alpha D_S), \lambda_k(V + \alpha D_S))| : j, k = 1, 2, ..., n\}.$$

Since $\sigma^2 = \|D_X\|^2 + \|D_S\|^2$, it suffices to prove the following inequality:

$$\max\{\varpi_1, \varpi_2\} \le \nu_2((\lambda_{\max}(V) + \alpha\sigma)^{p-1} + (\lambda_{\min}(V) - \alpha\sigma)^{-q-1}). \qquad (5.37)$$

From the choice of ϖ_1 we can safely claim that

$$\varpi_1 = |\Delta\psi'(\lambda_{j_*}(V + \alpha D_X), \lambda_{k_*}(V + \alpha D_X))|$$

for some index j_*, k_*. By the definition of $\Delta h'(\cdot, \cdot)$ and the mean value theorem [100], there exists a constant

$$\zeta_* \in [\min(\lambda_{j_*}(V + \alpha D_X), \lambda_{k_*}|(V + \alpha D_X)), \max(\lambda_{j_*}(V + \alpha D_X), \lambda_{k_*}(V + \alpha D_X))]$$

satisfying

$$\psi''(\zeta_*) = \Delta\psi'(\lambda_{j_*}(V + \alpha D_X), \lambda_{k_*}(V + \alpha D_X)).$$

This relation combined with Condition SR1 yields

$$\varpi_1 \le \nu_2\left(\zeta_*^{p-1} + \zeta_*^{-q-1}\right) \qquad (5.38)$$

for some

$$\zeta_* \in [\min(\lambda_i(V + \alpha D_X) : i \in I), \max(\lambda_i(V + \alpha D_X) : i \in I)].$$

Now let us recall the assumption in the lemma that $\alpha \in [0, \bar{\alpha})$, which further implies that, for any $i \in I$,

$$\lambda_{\min}(V) - \alpha\sigma \le \lambda_{\min}(V) - \alpha\|D_X\|_2 \le \lambda_i(V + \alpha D_X)$$

$$\le \lambda_{\max}(V) + \alpha\|D_X\|_2 \le \lambda_{\max}(V) + \alpha\sigma.$$

It follows immediately that

$$\lambda_{\min}(V) - \alpha\sigma \le \zeta_* \le \lambda_{\max}(V) + \alpha\sigma.$$

As a direct consequence of the above inequality and (5.38), we have

$$\varpi_1 \le \nu_2\left((\lambda_{\max}(V) + \alpha\sigma)^{p-1} + (\lambda_{\min}(V) - \alpha\sigma)^{-q-1}\right).$$

Similarly one can show

$$\varpi_2 \le \nu_2\left((\lambda_{\max}(V) + \alpha\sigma)^{p-1} + (\lambda_{\min}(V) - \alpha\sigma)^{-q-1}\right).$$

These two inequalities give (5.37), which further concludes the lemma. $\qquad \square$

Applying Lemma 5.2.8 to the function $g(\alpha)$ one easily gets

$$g(0) = g_1(0) = 0, \quad g'(0) = g_1'(0) = -\frac{\sigma^2}{2}.$$

Thus, from Lemma 5.4.2 we obtain

$$g(\alpha) \le g_1(\alpha) \le -\frac{\sigma^2 \alpha}{2} + \frac{\nu_2 \sigma^2}{2}$$

$$\times \int_0^\alpha \int_0^\xi \Big((\lambda_{\max}(V) + \zeta\sigma)^{p-1} + (\lambda_{\min}(V) - \zeta\sigma)^{-q-1} \Big) d\zeta d\xi,$$

which is essentially the same as its LO analogue (3.29) where the variables $v_{\max}, v_{\min}$ are replaced by $\lambda_{\max}(V), \lambda_{\min}(V)$, respectively. Let

$$g_2(\alpha) := -\frac{\sigma^2 \alpha}{2}$$

$$+ \frac{\nu_2 \sigma^2}{2} \int_0^\alpha \int_0^\xi \Big((\lambda_{\max}(V) + \zeta\sigma)^{p-1} + (\lambda_{\min}(V) - \zeta\sigma)^{-q-1} \Big) d\zeta d\xi.$$

Obviously, $g_2(\alpha)$ is convex and twice differentiable for all $\alpha \in [0, \bar{\alpha})$. Let α^* be the point at which $g_2(\alpha)$ attains its global minimal value, that is,

$$\alpha^* = \arg \min_{\alpha \in [0,\bar{\alpha})} g_2(\alpha) \tag{5.39}$$

as in (3.30). One can easily see that α^* is the unique solution of the following equation:

$$\frac{\nu_2}{p} \Big((\lambda_{\max}(V) + \alpha^* \sigma)^p - \lambda_{\max}(V)^p \Big) + \frac{\nu_2}{q} \Big((\lambda_{\min}(V) - \alpha^* \sigma)^{-q} - \lambda_{\min}(V)^{-q} \Big) = \sigma$$

$$\tag{5.40}$$

For this α^*, by means of Lemma 1.3.3, we obtain

$$g(\alpha^*) \le g_2(\alpha^*) \le \frac{1}{2} g_2'(0)\alpha^* = \frac{1}{2} g'(0)\alpha^*. \tag{5.41}$$

On the other hand, proceeding as in the proof of Lemma 3.3.3, one can get the following lower bound for α^*:

$$\alpha^* \ge \nu_5 \sigma^{-(q+1)/q},$$

where ν_5 is a positive constant similar to that defined in Lemma 3.3.3. Thus, if a step size $\alpha = \alpha^*$ or $\alpha = \nu_5 \sigma^{-(q+1)/q}$ is employed in the algorithm, then after one Newton step the following result holds.

Theorem 5.4.3 *Let the function $g(\alpha)$ be defined by (5.32) with $\Psi(V) \ge \nu_1^{-1}$. Then the step size $\alpha = \alpha^*$ defined by (5.39) or $\alpha = \nu_5 \sigma^{-(q+1)/q}$ is strictly feasible. Moreover,*

$$g(\alpha) \le \frac{1}{2} g'(0)\alpha \le -\frac{\nu_5 \nu_1^{(q-1)/(2q)}}{4} \Psi(V)^{(q-1)/(2q)}.$$

In the special case where $\psi(t) = \Upsilon_{p,q}(t)$ (2.5), this bound (with $\nu_1 = \nu_2 = 1$) simplifies to

$$g(\alpha) \leq -\min\left\{\frac{1}{12p+4}, \frac{1}{24q+16}\right\}\Psi(V)^{(q-1)/(2q)}.$$

Proof The proof is similar to its LO analogue. Note that from the definitions of $g(\alpha), g_1(\alpha)$ and $g_2(\alpha)$ for SDO, we obtain

$$g(\alpha) \leq g_1(\alpha) \leq g_2(\alpha) \leq -\frac{\nu_5}{4}\sigma^{(q-1)/q} \leq -\frac{\nu_5\nu_1^{(q-1)/(2q)}}{4}\Psi(V)^{(q-1)/(2q)},$$

where the second inequality is implied by (5.25) in Proposition 5.3.2. □

Since the proximity $\Psi(V)$ is determined by the eigenvalues of the matrix V, the growth behavior of the proximity $\Psi(V)$ is exactly the same as its LO counterpart $\Psi(v)$. If the current point is in the neighborhood $\mathcal{N}(\mu, \tau)$ of the central path, then we update μ to $(1 - \theta)\mu$ for some $\theta \in (0, 1)$. Proceeding as for the LO case, one can show that after the update of μ, the proximity is still bounded above by the number $\psi_0(\theta, \tau, n)$ defined by (3.40). We therefore immediately have the following bound.

Lemma 5.4.4 *Let $\Phi(X, S, \mu) \leq \tau$ and $\tau \geq \nu_1^{-1}$. Then, after an update of the barrier parameter, no more than*

$$\left\lceil \frac{4q\nu_1^{-(q-1)/(2q)}}{\nu_5}(\psi_0(\theta, \tau, n))^{(q+1)/(2q)} \right\rceil$$

iterations are needed to recenter. In the special case where $\psi(t) = \Upsilon_{p,q}(t)$ (2.5) (with $\nu_1 = \nu_2 = 1$), then after an update of the barrier parameter no more than

$$\left\lceil \frac{8q\max\{3p+1, 6q+4\}}{q+1}(\psi_0(\theta, \tau, n))^{(q+1)/(2q)} \right\rceil$$

inner iterations are needed to recenter.

Now we can state the final complexity of the algorithm.

Theorem 5.4.5 *If $\tau \geq \nu_1^{-1}$, the total number of iterations required by the primal-dual Newton algorithm is no more than*

$$\left\lceil \frac{4q\nu_1^{-(q-1)/(2q)}}{\nu_5}(\psi_0(\theta, \tau, n))^{(1+q)/(2q)} \right\rceil \left\lceil \frac{1}{\theta}\log\frac{n}{\varepsilon} \right\rceil.$$

In the special case where $\psi(t) = \Upsilon_{p,q}(t)$ (2.5) (with $\nu_1 = \nu_2 = 1$), the above bound for total iterations of the algorithm simplifies to

$$\left\lceil \frac{8q \max\{3p+1, 6q+4\}}{q+1} (\psi_0(\theta, \tau, n))^{(q+1)/(2q)} \right\rceil \left\lceil \frac{1}{\theta} \log \frac{n}{\varepsilon} \right\rceil.$$

Omitting the round off brackets in Theorem 5.4.5, one can conclude that by choosing $\theta \in (0, 1)$ and suitable $p, q \geq 1$, the complexity of our algorithm with large-update for SDO is an $\mathcal{O}(n^{(q+1)/(2q)} \log (n/\varepsilon))$ iterations bound, while the algorithm with small-update $(\theta = \mathcal{O}(1/\sqrt{n}))$ still has an $\mathcal{O}(\sqrt{n} \log(n/\varepsilon))$ iterations bound. Moreover, by means of Theorem 5.4.5, one can readily verify that if p is a constant and $q = \log n$, then the new large-update algorithm has an $\mathcal{O}(\sqrt{n} \log n \log(n/\varepsilon))$ bound for the total number of iterations.

Chapter 6

Primal-Dual Interior-Point Methods for Second-Order
Conic Optimization Based on Self-Regular Proximities

Based on the notion of self-regularity *associated with the second-order cone, this chapter deals with primal-dual Newton methods for solving SOCO problems. After a brief introduction to the problem under consideration, general analytical functions associated with the second-order cone are introduced and versatile properties of these functions are exploited. Special attention is paid to self-regular functions and self-regular proximities related to the second-order cone K. New search directions for large-update primal-dual IPMs for solving the underlying problem are then proposed and the complexity results of the corresponding algorithms are presented.*

6.1 INTRODUCTION TO SOCO, DUALITY THEORY AND THE CENTRAL PATH

Mathematically, a typical second-order cone can be defined by

$$K = \left\{ (x_1, x_2, \ldots, x_n) \in \Re^n : x_1^2 - \sum_{i=2}^{n} x_i^2 \geq 0, x_1 \geq 0 \right\}.$$

This cone is often referred to as the Lorentz cone in physics and verbally we also use its descriptive nickname: the ice-cream cone.[1] Second-order conic optimization (SOCO) is the problem of minimizing a linear objective function subject to the intersection of an affine set and the direct product of several second-order cones. Hence, from a pure mathematical viewpoint, the constraint functions defining the second-order cone are nothing more than some specific quadratic functions.

In light of the above-mentioned relation, SOCO is always recognized as a generalization of LO. Several important types of problems can be modeled as SOCO problems. For example, a general convex quadratic optimization problem with convex quadratic constraints can be cast as a SOCO problem [81]. Also included are specific cases: robust linear optimization, robust least-squares, matrix-fractional problems, and problems involving sums and maxima of norms, etc. SOCO has been widely applied in several areas for a long time. One can list, for example, antenna array weight design, grasping force optimization, FIR filter design, portfolio optimization with loss risk constraints, truss design, etc. For more details about different applications of SOCO, refer to the survey paper [63] and the references therein.

An alternative way to describe the second-order cone is via a matrix expression. For any $x = (x_1, \ldots, x_n)^T \in \Re^n$, let us define the matrix

$$\text{mat}(x) = \begin{pmatrix} x_1 & x_{2:n} \\ x_{2:n}^T & x_1 E_{n-1} \end{pmatrix} \tag{6.1}$$

where $x_{2:n} = (x_2, x_3, \ldots, x_n)$. With this definition, one can easily prove that the vector $x \in K$ if and only if the matrix $\text{mat}(x)$ is positive semidefinite, that is, $\text{mat}(x) \succeq 0$. This observation means that SOCO is essentially a specific case of SDO. Thus, one has the "sandwich" relation LO $\subset$ SOCO $\subset$ SDO. This delicate circumstance explains partially why SOCO did not attract as much attention as its counterparts LO and SDO in the surge of study on IPMs.

Recently this situation started to change gradually when it was observed that (as pointed out in [83]) although a SOCO problem can be solved by using an SDO approach, IPMs that solve the SOCO problem directly have much

---

[1] This interesting name comes from the similarity between the shape of a general second-order cone in the space $\Re^3$ and the enjoyable summer food.

better complexity than an IPM applied to the semidefinite formulation of the SOCO problem. The reason behind this observation is clear: IPMs with small-update (or large-update) working directly on SOCO have polynomial $\mathcal{O}(\sqrt{N}\log((x^0)^{\mathrm{T}}s^0/(n\varepsilon)))$ (or $\mathcal{O}(N\log((x^0)^{\mathrm{T}}s^0/(n\varepsilon))))$ iteration bounds, where N is the number of cones involved in the underlying problem and n is number of variables in the matrix, while for the SDO algorithm, the iteration number is bounded above by $\mathcal{O}(\sqrt{n}\log((x^0)^{\mathrm{T}}s^0/(n\varepsilon)))$ (or $\mathcal{O}(n\log((x^0)^{\mathrm{T}}s^0/(n\varepsilon))))$ and in general n is a much bigger number than N. Another important observation is that in practice, at each iteration much less work is needed for the IPMs based on SOCO. Moreover, even if the same type of search direction is employed, say the AHO (see our discussion in Section 6.3) direction, the theoretical properties of the corresponding algorithm for SOCO can be enhanced slightly compared with its analogue for SDO [78]. These theoretical conclusions are further strengthened by recent extensive numerical experiments on SOCO [8], where several SOCO problems with hundreds of thousands of variables are reported to be solved efficiently. On the other hand, we would like to mention again that, although IPMs provide a powerful approach for SDO, to solve a general SDO of thousands of variables without any special data structure is still a serious challenge [120].

In this chapter, we consider the standard SOCO problem, which takes the following form:

$$(\text{SOCO}) \quad \min c^{\mathrm{T}}x$$

$$Ax = b, x \succeq_K 0,$$

and its dual

$$(\text{SOCD}) \quad \max b^{\mathrm{T}}y$$

$$A^{\mathrm{T}}y + s = c, s \succeq_K 0,$$

where K is the product of several second-order cones, that is, $K = K^1 \times K^2 \times \cdots \times K^N$ with

$$K^j = \left\{ (x_1^j, ..., x_{n_j}^j)^{\mathrm{T}} \in \mathfrak{R}^{n_j} : \left(x_1^j\right)^2 \geq \sum_{i=2}^{n_j} \left(x_i^j\right)^2, \ x_1^j \geq 0 \right\},$$

and K_+ is the interior region of K, $A \in \mathfrak{R}^{m \times n}$ with $n = \sum_{j=1}^N n_j$, and

$$x = \begin{pmatrix} x^1 \\ x^2 \\ \vdots \\ x^N \end{pmatrix}, \quad x^j \in \mathfrak{R}^{n_j}, \quad j \in J, \quad x \in \mathfrak{R}^n,$$

where J denotes the index set $J=\{1,2,...,N\}$. As standard, the notation $x \succeq_K s$ (or $x \succ_K s$) means that $x - s \in K$ (or $x - s \in K_+$). In this chapter the matrix A is further assumed to be of full row rank, that is, rank $(A) = m$.

Because of the inherent relations among LO, SOCO and SDO, most theoretical results for SOCO can be viewed as a transparent extension of LO or a specialization of the results for SDO. For example, the duality theory for SOCO and SDO has indeed been built up on a unified conic form [13,83, 103]. Hence, we have directly the well-known weak duality theory: if x is feasible for the primal (SOCO) and (y, s) is feasible for the dual (SOCD), then $b^T y \le c^T x$. Regarding the strong duality for SOCO, one has the following results [13,83,103].

Theorem 6.1.1 *If both of the following conditions hold:*

(i) (SOCO) is strictly feasible, that is, there exists an $x \in \Re^n$ such that

$$Ax = b, \quad x \succ_K 0;$$

(ii) (SOCD) is strictly feasible, that is, there exists $s \in \Re^n, y \in \Re^m$ such that

$$A^T y + s = c, \quad s \succ_K 0;$$

then we have a pair of optimal solutions x^*, (y^*, s^*) with $c^T x^* = b^T y^*$. Further, if condition (i) holds and the primal objective is bounded from below, then there is a dual optimal solution (y^*, s^*) of (SOCD) satisfying

$$b^T y^* = \inf\{c^T x : Ax = b, \ x \succeq_K 0\};$$

if condition (ii) holds and the dual objective is bounded from above, then there is a primal optimal solution x^* of (SOCO) satisfying

$$c^T x^* = \sup\{b^T y : A^T y + s = c, \ s \succeq_K 0\}.$$

Here $x \succ_K 0$ means that x is in the interior region of K. For algorithmic purposes, we assume in this chapter that both the primal problem and its dual are strictly feasible. Thus, by Theorem 6.1.1 one can claim that if (x^*, y^*, s^*) is an optimal solution of the primal-dual pair (SOCO) and (SOCD), then

$$(x^*)^T s^* = c^T x^* - b^T y^* = 0,$$

which, combined with the assumption $x^*, s^* \succeq_K 0$, gives

$$\mathrm{mat}(x^*)s^* = 0.$$

Here $\mathrm{mat}(x) = \mathrm{diag}(\mathrm{mat}\,(x^1), ..., \mathrm{mat}\,(x^N))$ is the notation introduced by (6.1).

Akin to its LO and SDO analogue, the central path for SOCO is defined by the solution sets $\{x(\mu), y(\mu), s(\mu) : \mu > 0\}$ of the following system:

$$\begin{cases} Ax & = b, \\ A^T y + s & = c, \\ \text{mat}(xs) & = \mu\tilde{e}, \quad x, s \succeq_K 0, \end{cases} \tag{6.2}$$

where

$$\tilde{e} = \begin{pmatrix} \tilde{e}^1 \\ \tilde{e}^2 \\ \vdots \\ \tilde{e}^N \end{pmatrix}, \quad \tilde{e}^j = \begin{pmatrix} 1 \\ 0 \\ \vdots \\ 0 \end{pmatrix} \in \Re^{n_j}, \quad j \in J.$$

The next result dealing with the uniqueness and existence of the central path is due to Nesterov and Nemirovskii [83]; see also [120].

Theorem 6.1.2 *Suppose that both (SOCO) and (SOCD) are strictly feasible. Then, for every positive μ, there is a unique solution $(x(\mu), y(\mu), s(\mu))$ in $K \times \Re^m \times K$ to the system (6.2).*

The central path behaves in a very similar way to the central path for SDO. Thus we omit the discussion about the limiting behavior of such a path and refer the reader to references [31,55,103,120] for details. IPMs trace the central path appropriately and find an approximate solution to the SOCO problem in question as μ decreases to zero.

6.2 PRELIMINARY RESULTS ON FUNCTIONS ASSOCIATED WITH SECOND-ORDER CONES

As we have seen in the previous chapters, to define the new search direction and analyze the complexity of the algorithm, we always resort to some functions defined in a suitable space. In this section we first give several fundamental results about general functions associated with the second-order cone. We note that in [19], using Jordan algebra, the authors developed some deep and abstract theory for various functions in the so-called v-space. However, the results presented in [19] are not easily understandable and cannot be applied to IPMs directly. Later on Faybusovich [20,21] recognized the significance of Euclidean Jordan algebra for second-order cones and used it to analyze IPMs. With regard to the algorithmic aspect, Faybusovich's results rely, more or less, on the logarithmic barrier approach that originated from the book [83] or the papers [84,86]. Tsuchiya [113] and Monteiro and Tsuchiya [78] then applied Jordan algebra to analyze IPMs for SOCO with specialization in various search directions.

In this section, following the aforementioned approach, we study general functions associated with the second-order cone via Jordan algebra. To ease the discussion, in this technical section we assume the cone K is defined with $N = 1$. First we observe that, closely associated with the cone K is a matrix

$$Q = \text{diag}(1, -1, ..., -1).$$

We refer to Q as the representation matrix of the cone K, since the definition of K is equivalent to

$$K = \{x \in \Re^n : x^T Q x \geq 0, \ x_1 \geq 0\}.$$

Obviously one has $Q^2 = E$.

6.2.1 Jordan Algebra, Trace and Determinant

In this subsection we first describe the so-called Jordan algebra related to the second-order cone and then present some technical conclusions about the trace and determinant of vectors relevant to the second-order cone.

The Euclidean Jordan algebra for the second-order cone K is defined by the bilinear operator:

$$x \circ s = (x^T s, x_1 s_2 + s_1 x_2, ..., x_1 s_n + s_1 x_n)^T = (x^T s, x_1 s_{2:n}^T + s_1 x_{2:n}^T)^T. \quad (6.3)$$

It is easy to verify that

$$x \circ s = \text{mat}(x)s = \text{mat}(s)x = s \circ x.$$

Obviously, the Jordan product $\circ$ is commutative. The cone K is not closed under the Jordan product. For example, if $n = 3$, then $x = (1.5, 1, 1)^T \in K$ and $s = (1.5, 1, -1)^T \in K$, but $x \circ s = (2.25, 3, 0)^T \notin K$. Another interesting and important fact, as easily observed, is that for any $z \in \Re^n$ one has $z \circ z \in K$. In other words, the Jordan square of any vector belongs to K. On the other hand, for every nonzero $x \in K$, the equation $z \circ z = x$ has two solutions z with $z_1 \geq 0$, namely

$$z_1 = \sqrt{\frac{1}{2}(x_1 + \lambda)}, \quad z_i = \frac{x_i}{2z_1}, \quad 2 \leq i \leq n, \quad (6.4)$$

where

$$\lambda = \pm \sqrt{x_1^2 - \sum_{i \geq 2} x_i^2}.$$

One may easily verify that $z \in K$ if $\lambda \geq 0$, whereas $z \notin K$ if $\lambda < 0$. We conclude from this that for every $x \in K$, the equation $z \circ z = x$ has a unique solution z in K. Later on we denote this solution as $x^{1/2}$, or as $\sqrt{x}$, and we consider more general functions on K. Before doing so, we need to introduce the concepts of trace and determinant with respect to K.

Let us denote by $\lambda_{\max}(x)$ and $\lambda_{\min}(x)$ the maximal and minimal eigenvalues of the matrix $\mathrm{mat}(x)$, respectively. Namely

$$\lambda_{\max}(x) = x_1 + \|x_{2:n}\|, \quad \lambda_{\min}(x) = x_1 - \|x_{2:n}\|. \tag{6.5}$$

The trace and determinant of a vector $x \in \Re^n$ associated with the second-order cone K are defined as follows.

Definition 6.2.1 *For any $x \in \Re^n$, the trace of x is defined by*

$$\mathrm{Tr}(x) = \lambda_{\max}(x) + \lambda_{\min}(x) = 2x_1 \tag{6.6}$$

and the determinant of x is given by

$$\det(x) = x_1^2 - \|x_{2:n}\|^2 = \lambda_{\max}(x)\lambda_{\min}(x).$$

These definitions can be viewed as variants of the determinant and trace of general matrices. From these definitions, one can easily see that for any $x \in \Re^n$,

$$\mathrm{Tr}(x \circ x) = 2\|x\|^2.$$

Moreover, it is straightforward to verify the following result.

Lemma 6.2.2 *Suppose the trace and determinant of a vector $x \in \Re^n$ are defined by (6.6) and (6.7), respectively. Then $x \succeq_K 0$ (or $x \succ_K 0$) if and only if $\mathrm{Tr}(x) \geq 0$ and $\det(x) \geq 0$ (or $\mathrm{Tr}(x) > 0$ and $\det(x) > 0$).*

Our next lemma collects several elementary results about the behavior of the trace and determinant of the Jordan product of two vectors. These results illustrate the difference between the elements in the second-order cone K and the cone of positive definite matrices.[2]

Lemma 6.2.3 *Suppose that x and s are two vectors in $\Re^n$. Then we have*

$$\lambda_{\max}(x)\lambda_{\min}(s) + \lambda_{\min}(x)\lambda_{\max}(s) \leq \mathrm{Tr}(x \circ s), \tag{6.8}$$

$$\mathrm{Tr}(x \circ s) \leq \lambda_{\max}(x)\lambda_{\max}(s) + \lambda_{\min}(x)\lambda_{\min}(s) \tag{6.9}$$

and

$$\det(x \circ s) \leq \det(x)\det(s). \tag{6.10}$$

Furthermore, the equality in (6.10) holds if and only if the two vectors $x_{2:n}$ and $s_{2:n}$ are linearly dependent.

---

[2] This distinction can be expected by noticing that $\mathrm{Tr}(\mathrm{mat}(x)) = nx_1 = \mathrm{Tr}(x)n/2$ and that

$$\det(\mathrm{mat}(x)) = x_1^{n-2}(x_1^2 - \|x_{2:n}\|^2) = x_1^{n-2}\det(x).$$

Proof We first consider inequalities (6.8) and (6.9). Using the notation $\lambda_{\max}(\cdot)$ and $\lambda_{\min}(\cdot)$ given by (6.5) and the Cauchy–Schwarz inequality, one has

$$\lambda_{\max}(x)\lambda_{\min}(s) + \lambda_{\min}(x)\lambda_{\max}(s) = 2(x_1 s_1 - \|x_{2:n}\| \, \|s_{2:n}\|)$$

$$\leq 2x^{\mathrm{T}}s = \mathrm{Tr}(x \circ s) \leq 2(x_1 s_1 + \|x_{2:n}\| \, \|s_{2:n}\|)$$

$$= \lambda_{\max}(x)\lambda_{\max}(s) + \lambda_{\min}(x)\lambda_{\min}(s),$$

which gives (6.8) and (6.9).

To prove (6.10), we note that by making use of definition (6.7), one gets

$$\det(x \circ s) = (x^{\mathrm{T}}s)^2 - \|x_1 s_{2:n}^{\mathrm{T}} + s_1 x_{2:n}^{\mathrm{T}}\|^2$$

$$= (x_1 s_1)^2 + (x_{2:n}^{\mathrm{T}} s_{2:n})^2 - (x_1)^2\|s_{2:n}\|^2 - (s_1)^2\|x_{2:n}\|^2$$

$$\leq (x_1 s_1)^2 + \|x_{2:n}\|^2\|s_{2:n}\|^2 - (x_1)^2\|s_{2:n}\|^2 - (s_1)^2\|x_{2:n}\|^2$$

$$= \left((x_1)^2 - \|x_{2:n}\|^2\right)\left((s_1)^2 - \|s_{2:n}\|^2\right) = \det(x)\det(s);$$

and the equality holds if and only if $|x_{2:n}^{\mathrm{T}} s_{2:n}| = \|x_{2:n}\| \, \|s_{2:n}\|$. This means the equality holds only when the vectors $x_{2:n}$ and $s_{2:n}$ are linearly dependent. $\square$

6.2.2 Functions and Derivatives Associated with Second-Order Cones

We now pose a new definition of general functions associated with the second-order cone K and study various properties of these functions. As shown by Fukushima, Luo and Tseng [27], this is a far from trivial task. Note that if $n = 1$, then $K = \Re_+$, and if $n \geq 1$, then $\Re_+ \subseteq K$. In this case, a function on the cone has the same form as that of a standard function in $\Re$ or $\Re_+$. In the sequel we show that any function mapping $\Re_+$ into $\Re_+$ can be naturally extended to a function that maps K into itself. We start with the basic definition with respect to general functions associated with the second-order cone K.

Definition 6.2.4[3] *Suppose that $\psi(t)$ is a function from $\Re$ to $\Re$, and $x \in \Re^n$. Then the function $\psi(x) : \Re^n \to \Re^n$ associated with the second-order cone K is defined as follows:*

$$\psi(x) = \left(\frac{1}{2}(\psi(\lambda_{\max}(x)) + \psi(\lambda_{\min}(x))), \Delta\psi(\lambda_{\max}(x), \lambda_{\min}(x))x_{2:n}^{\mathrm{T}}\right)^{\mathrm{T}} \quad (6.11)$$

---

[3] We notice that in [27], the authors posed a slightly different definition of functions associated with a second-order cone. However, our definition (6.11) is clearer and more direct.

In the special case $x_{2:n} = 0$, we define

$$\psi(x) := (\psi(\lambda_{max}(x)), 0, ..., 0)^T.$$

It can easily be verified that if $\psi(t) \geq 0$ for any $t \geq 0$ and $x \in K$, then the above definition implies that $\psi(x) \in K$. Thus, it becomes clear that every nonnegative (positive) function on the nonnegative (positive) axis naturally extends to a function that maps (the interior of) K into itself. Likewise for the LO and SDO cases, the function $\psi(t)$ is called the kernel function of $\psi(x)$. As a consequence of the above definition we have a large source of functions mapping K into itself. For instance, we may write x^p, where p is any number in $\Re$ and $x \in K$. Let us consider some concrete choices; for example, $p = -1$. In such a situation definition (6.11) yields

$$x^{-1} = \frac{1}{\det(x)}(x_1, -x_2, -x_3, ..., -x_n)^T \quad \forall x \succ_K,$$

and one may easily see that $x \circ x^{-1} = \tilde{e}$. Let us consider another case where $p = 1/2$, then $\sqrt{x}$, as defined by definition (6.11) is given by (6.4). Moreover, it is clear that any analytical functions like $\exp(x)$, or $\sin(x)$ are now well defined. Similarly we can define the function $\psi'(x)$ by (6.11) whose kernel function is $\psi'(t)$.

However, this raises a question. For example, with $\psi(t) = t^2$, can we justify writing $\psi(x) = x^2$? The answer is affirmative due to the fact that in that case $\psi(x) = x \circ x$, as can be easily verified. More generally, when functions are composed of products, the behavior with respect to the Jordan product is nice, as the following lemma shows.

Lemma 6.2.5 *Suppose that $\psi_1(t)$, $\psi_2(t)$ are two functions from $\Re_{++}$ into $\Re$ and $\psi_1(x), \psi_2(x)$ are two associated functions defined by (6.11). If $\psi_3(t) = \psi_1(t)\psi_2(t)$, then $\psi_3(x) = \psi_1(x) \circ \psi_2(x)$.*

Proof We first consider the case where $\|x_{2:n}\| > 0$. From (6.11) we deduce

$$\psi_j(x) = \left(\frac{1}{2}(\psi_j(\lambda_{max}(x)) + \psi_j(\lambda_{min}(x))), \frac{\psi_j(\lambda_{max}(x)) - \psi_j(\lambda_{min}(x))}{2\|x_{2:n}\|}x_{2:n}\right)^T$$

for $j = 1, 2, 3$. Hence, by simple computation, one obtains

$$(\psi_1(x) \circ \psi_2(x))_1 = \frac{1}{2}(\psi_1(\lambda_{max}(x))\psi_2(\lambda_{max}(x)) + \psi_1(\lambda_{min}(x))\psi_2(\lambda_{min}(x)))$$

and

$$(\psi_1(x) \circ \psi_2(x))_i = \frac{x_i}{2\|x_{2:n}\|}(\psi_1(\lambda_{max}(x))\psi_2(\lambda_{max}(x)) - \psi_1(\lambda_{min}(x))\psi_2(\lambda_{min}(x))),$$

for any $i \in \{2, ..., n\}$. The statement of the lemma follows immediately from these two relations. It is straightforward to verify the conclusion of the lemma when $\|x_{2:n}\| = 0$. $\qquad\qquad\qquad\qquad\qquad\qquad\qquad\qquad\qquad\qquad\qquad$ □

It is trivial to verify the following result about common functions associated with the second-order cone defined by (6.11).

Lemma 6.2.6 *Let the associated function $\psi(x) : \Re^n \to \Re^n$ be defined by (6.11). Then*

$$\|\psi(x)\| = \frac{\sqrt{2}}{2}\sqrt{\psi^2(\lambda_{\max}(x)) + \psi^2(\lambda_{\min}(x))},$$

$$\mathrm{Tr}\,(\psi(x)) = \psi(\lambda_{\max}(x)) + \psi(\lambda_{\min}(x)),$$

$$\det\,(\psi(x)) = \psi(\lambda_{\max}(x))\psi(\lambda_{\min}(x)).$$

Now we start to explore the relations among the eigenvalues of two vectors and the eigenvalues of the Jordan product based on them. Note from Lemma 6.2.3 that for any $x, s \succ_K 0$, the determinant $\det(x \circ s) = \det(x)\det(s)$ if and only if the vectors $x_{2:n}$ and $s_{2:n}$ are linearly dependent. Without loss of generality, we may assume that $x_{2:n} \neq 0$ and $s_{2:n} = \beta x_{2:n}$ for some $\beta \in \Re$. In the sequel we show this further implies that the vector s can be represented as a function of x and that the vector $x \circ s \in K_+$.

Lemma 6.2.7 *Suppose that x and s are two vectors belonging to K_+ with $x_{2:n} \neq 0$. If*

$$\det(x \circ s) = \det(x)\det(s),$$

then there exists a function $\psi(t) : \Re_{++} \to \Re_{++}$ such that $s = \psi(x)$ and the Jordan product $x \circ s \in K_+$. Moreover,

$$\lambda_{\min}(x)\lambda_{\min}(s) \le \lambda_{\min}(x \circ s) \le \lambda_{\max}(x \circ s) \le \lambda_{\max}(x)\lambda_{\max}(s). \qquad (6.12)$$

Proof By making use of the last conclusion of Lemma 6.2.3, since $\det(x \circ s) = \det(x)\det(s)$, one can claim that

$$s_{2:n} = \frac{\beta\|s_{2:n}\|}{\|x_{2:n}\|}x_{2:n},$$

where β is 1 or -1. Let $\psi(t)$ be a univariate function from $\Re_{++}$ into $\Re_{++}$ satisfying

$$\frac{1}{2}(\psi(\lambda_{\max}(x)) + \psi(\lambda_{\min}(x))) = s_1, \quad \frac{1}{2}(\psi(\lambda_{\max}(x)) - \psi(\lambda_{\min}(x))) = \beta\|s_{2:n}\|.$$

Such a function exists because $s \in K_+$. Thus, by definition (6.11) we can claim $s = \psi(x)$. Now invoking Lemma 6.2.5, we can write $x \circ s = \psi_1(x)$, where the kernel function $\psi_1(t) = t\psi(t)$. By the choice of $\psi(t)$ we know that $\psi_1(t) > 0$ for any $t > 0$. Therefore, from definition (6.11) it follows that $x \circ s = \psi_1(x) \succ_K 0$.

It remains to prove (6.12). Since $x \circ s = \psi_1(x)$, we have

$$\lambda_{\max}(x \circ s) = \max\{\psi_1(\lambda_{\max}(x)), \psi_1(\lambda_{\min}(x))\}$$

$$\leq \lambda_{\max}(x) \max\{\psi(\lambda_{\max}(x)), \psi(\lambda_{\min}(x))\} = \lambda_{\max}(x)\lambda_{\max}(s),$$

which, along with the assumption in the lemma that $\det(x \circ s) = \det(x)\det(s)$, further yields

$$\lambda_{\min}(x \circ s) \geq \lambda_{\min}(x)\lambda_{\min}(s).$$

The proof of the lemma is completed. $\square$

We remark that the results in the above lemma can be slightly strengthened, since for any $x, s \in K_+$ one can easily see that

$$\lambda_{\max}(x \circ s) = x^T s + \|x_1 s_{2:n} + s_1 x_{2:n}\| \leq (x_1 + \|x_{2:n}\|)(s_1 + \|s_{2:n}\|)$$

$$= \lambda_{\max}(x)\lambda_{\max}(s).$$

However, it seems impossible to prove the relation

$$\lambda_{\min}(x \circ s) \geq \lambda_{\min}(x)\lambda_{\min}(s)$$

without the condition imposed in Lemma 6.2.7. This is because, for general $x, s \in K_+$, if $\det(x \circ s) \neq \det(x)\det(s)$, then the vector $x \circ s$ might be on the boundary of K or outside K. For instance, let $x = (2, 1, 1)^T$, $s = (2, 1, -1)$. It is straightforward to see that

$$\lambda_{\min}(x) = \lambda_{\min}(s) = 2 - \sqrt{2} > 0.$$

On the other hand, one can readily verify that $\lambda_{\min}(x \circ s) = 0$.

Let us recall some known inequalities in matrix theory. Suppose that X and S are both symmetric and positive definite. Let $\lambda_{\max}(\cdot)$ and $\lambda_{\min}(\cdot)$ denote the maximal and minimal eigenvalues of the corresponding matrix. Then we have[4]

---

[4] It suffices to prove relation (6.14), since inequality (6.13) follows from (6.14) applied to the inverse matrices X^{-1} and S^{-1}. Note that, by using relation (3.3.20) in [40, pp. 178], we get

$$\lambda_{\max}(XS) \leq \varrho_{\max}(XS) \leq \lambda_{\max}(X)\lambda_{\max}(S),$$

where $\varrho_{\max}(XS)$ denotes the maximal singular value of XS.

$$\lambda_{\min}(X)\lambda_{\min}(S) \le \lambda_{\min}(XS), \tag{6.13}$$

$$\lambda_{\max}(XS) \le \lambda_{\max}(X)\lambda_{\max}(S). \tag{6.14}$$

Lemma 6.2.7 resembles relations (6.13) and (6.14) in the cone of symmetric positive definite matrices. Note that for any $X, S \in \Re^{n \times n}$, there holds trivially $\det(XS) = \det(X)\det(S)$. The results presented in Lemma 6.2.7 are very helpful in our later discussion about the features of a self-regular function associated with the second-order cone whose definition is recorded as follows. This definition coincides with its SDO cousin in the sense that both of them are derived from a univariate self-regular kernel function $\psi(t)$.

Definition 6.2.8   *A function $\psi(x)$ associated with the second-order cone K given by (6.11) is said to be self-regular if its kernel function $\psi(t)$ is self-regular.*

As for its SDO analogue, we denote by $\Psi(x)$ the trace of the function $\psi(x)$, that is,

$$\Psi(x) = \text{Tr}(\psi(x)) = \psi(\lambda_{\max}(x)) + \psi(\lambda_{\min}(x)). \tag{6.15}$$

Our next proposition characterizes several important properties of a self-regular function associated with the second-order cone K.

Proposition 6.2.9   *Let the functions $\psi(x) : K_+ \to K_+$ and $\Psi(x) : K_+ \to \Re$ be defined by (6.11) and (6.15), respectively. If the function $\psi(x)$ is self-regular, then the following statements hold:*

(i) *$\Psi(x)$ is strictly convex with respect to $x \in K_+$ and vanishes at its global minimal point $x = \tilde{e}$, that is, $\Psi(\tilde{e}) = 0$, $\psi(\tilde{e}) = \psi'(\tilde{e}) = 0$. Further there exist positive constants $\nu_1, \nu_2 > 0$ and $p, q \ge 1$ such that*

$$\nu_1(x^{p-1} + x^{-1-q}) \preceq_K \psi''(x) \preceq_K \nu_2(x^{p-1} + x^{-1-q}). \tag{6.16}$$

(ii) *Suppose x and s are two vectors in K_+. If $v \in K_+$ satisfies*

$$\det(v^2) = \det(x)\det(s), \quad \text{Tr}(v^2) = \text{Tr}(x \circ s),$$

   then

$$\Psi(v) \le \frac{1}{2}(\Psi(x) + \Psi(s)).$$

Proof To prove that the first claim of the proposition, we need to show that $\Psi(x)$ is strictly convex for $x \succ_K 0$; that is, for any $x, s \succ_K 0$ and $x \ne s$,

$$\Psi\left(\frac{x+s}{2}\right) < \frac{1}{2}(\Psi(x) + \Psi(s)).$$

Since $x, s \in K_+$, using simple calculus one can prove that

$$\lambda_{\max}\left(\frac{x+s}{2}\right) = \frac{x_1 + s_1}{2} + \frac{1}{2}\|x_{2:n} + s_{2:n}\| \leq \frac{1}{2}(\lambda_{\max}(x) + \lambda_{\max}(s)),$$

and similarly

$$\lambda_{\min}\left(\frac{x+s}{2}\right) = \frac{x_1 + s_1}{2} - \frac{1}{2}\|x_{2:n} + s_{2:n}\| \geq \frac{1}{2}(\lambda_{\min}(x) + \lambda_{\min}(s)).$$

Recalling the definitions of $\lambda_{\max}(\cdot)$ and $\lambda_{\min}(\cdot)$, we trivially have

$$\lambda_{\max}\left(\frac{x+s}{2}\right) + \lambda_{\min}\frac{x+s}{2} = \frac{1}{2}(\lambda_{\max}(x) + \lambda_{\min}(x) + \lambda_{\max}(s) + \lambda_{\min}(s)).$$

Thus, from the above three relations we can conclude that there exist two constants $\beta_1 \geq 0$ and $\beta_2 \geq 0$ with $\beta_1 + \beta_2 = 1$ such that

$$\lambda_{\max}\left(\frac{x+s}{2}\right) = \frac{\beta_1}{2}(\lambda_{\min}(x) + \lambda_{\min}(s)) + \frac{\beta_2}{2}(\lambda_{\max}(x) + \lambda_{\max}(s)),$$

$$\lambda_{\min}\left(\frac{x+s}{2}\right) = \frac{\beta_2}{2}(\lambda_{\min}(x) + \lambda_{\min}(s)) + \frac{\beta_1}{2}(\lambda_{\max}(x) + \lambda_{\max}(s)).$$

Now, making use of the strict convexity of the function $\psi(t)$ twice, one obtains

$$\Psi\left(\frac{x+s}{2}\right) = \psi\left(\lambda_{\max}\left(\frac{x+s}{2}\right)\right) + \psi\left(\lambda_{\min}\left(\frac{x+s}{2}\right)\right)$$

$$= \psi\left(\frac{\beta_1}{2}(\lambda_{\min}(x) + \lambda_{\min}(s)) + \frac{\beta_2}{2}(\lambda_{\max}(x) + \lambda_{\max}(s))\right)$$

$$+ \psi\left(\frac{\beta_2}{2}(\lambda_{\min}(x) + \lambda_{\min}(s)) + \frac{\beta_1}{2}(\lambda_{\max}(x) + \lambda_{\max}(s))\right)$$

$$\leq \psi\left(\frac{\lambda_{\min}(x) + \lambda_{\min}(s)}{2}\right) + \psi\left(\frac{\lambda_{\max}(x) + \lambda_{\max}(s)}{2}\right)$$

$$\leq \frac{1}{2}(\psi(\lambda_{\max}(x)) + \psi(\lambda_{\min}(x)) + \psi(\lambda_{\max}(s)) + \psi(\lambda_{\min}(s)))$$

$$= \frac{1}{2}(\Psi(x) + \Psi(s)).$$

Note that since $x \neq s$, at least one of the two inequalities in the above proof holds strictly. This proves the strict convexity of $\Psi(x)$. The remaining terms in the first statement can be verified using direct calculus.

It remains to prove the second statement of the proposition. For this we first observe that, since $v \in K_+$,

$$\det(v) = \det(v^2)^{1/2} = \det(x)^{1/2} \det(s)^{1/2}$$

$$= (\lambda_{\min}(x)\lambda_{\min}(s))^{1/2}(\lambda_{\max}(x)\lambda_{\max}(s))^{1/2} \qquad (6.18)$$

and

$$\text{Tr}(v) = \lambda_{\max}(v) + \lambda_{\min}(v) = (\lambda_{\max}(v)^2 + \lambda_{\min}(v)^2 + 2\lambda_{\max}(v)\lambda_{\min}(v))^{1/2}$$

$$= (\text{Tr}(v)^2 + 2\det(v))^{1/2} = \left(\text{Tr}(x \circ s) + 2(\det(x)\det(s))^{1/2}\right)^{1/2}$$

$$\le \left(\lambda_{\min}(x)\lambda_{\min}(s) + \lambda_{\max}(x)\lambda_{\max}(s) + 2(\det(x)\det(s))^{1/2}\right)^{1/2}$$

$$= \sqrt{\lambda_{\min}(x)\lambda_{\min}(s)} + \sqrt{\lambda_{\max}(x)\lambda_{\max}(s)}, \qquad (6.19)$$

where the inequality follows from (6.9). Therefore,

$$\lambda_{\max}(v) = \frac{1}{2}(\text{Tr}(v) + \lambda_{\max}(v) - \lambda_{\min}(v)) = \frac{1}{2}\text{Tr}(v) + \frac{1}{2}\sqrt{\text{Tr}(v)^2 - 4\det(v)^2}$$

$$\le \frac{1}{2}\text{Tr}(v) + \frac{1}{2}\left(\sqrt{\lambda_{\max}(x)\lambda_{\max}(s)} - \sqrt{\lambda_{\min}(x)\lambda_{\min}(s)}\right)$$

$$\le \lambda_{\max}(x)^{1/2}\lambda_{\max}(s)^{1/2},$$

where all inequalities follow from (6.18) and (6.19). Now, invoking (6.18), we can further claim

$$\lambda_{\min}(v) \ge \lambda_{\min}(x)^{1/2}\lambda_{\min}(s)^{1/2}.$$

From the above discussions we can easily verify that there exists a constant $r \in [1/2, 1)$ such that

$$\lambda_{\min}(v) = \lambda_{\min}(x)^{r/2}\lambda_{\min}(s)^{r/2}\lambda_{\max}(x)^{(1-r)/2}\lambda_{\max}(s)^{(1-r)/2},$$

$$\lambda_{\max}(v) = \lambda_{\max}(x)^{r/2}\lambda_{\max}(s)^{r/2}\lambda_{\min}(x)^{(1-r)/2}\lambda_{\min}(s)^{(1-r)/2}.$$

By applying Condition SR2 twice, we deduce

$$\Psi(v) = \psi(\lambda_{\min}(v)) + \psi(\lambda_{\max}(v))$$

$$= \psi\left(\lambda_{\min}(x)^{r/2}\lambda_{\min}(s)^{r/2}\lambda_{\max}(x)^{(1-r)/2}\lambda_{\max}(s)^{(1-r)/2}\right)$$

$$+ \psi\left(\lambda_{\max}(x)^{r/2}\lambda_{\max}(s)^{r/2}\lambda_{\min}(x)^{(1-r)/2}\lambda_{\min}(s)^{(1-r)/2}\right)$$

$$\le \psi\left(\lambda_{\min}(x)^{1/2}\lambda_{\min}(s)^{1/2}\right) + \psi\left(\lambda_{\max}(x)^{1/2}\lambda_{\max}(s)^{1/2}\right)$$

$$\leq \frac{1}{2}(\psi(\lambda_{\min}(x)) + \psi(\lambda_{\max}(x)) + \psi(\lambda_{\max}(s)) + \psi(\lambda_{\min}(s)))$$

$$= \frac{1}{2}(\Psi(x) + \Psi(s)).$$

This completes the proof of the proposition. $\qquad\qquad\qquad\square$

Comparing Proposition 6.2.9 with its SDO analogue Proposition 5.2.6, we find that statements (ii) in these two propositions are slightly different. Actually, one can easily see that the matrix used in the second claim of Proposition 5.2.6 satisfies certain conditions similar to those posed in Proposition 6.2.9. However the choice of the vector v allowing such conditions in second-order cones is much more strict. One possible reason for this phenomenon is that, for general $x, s \in K_+$, the Jordan product $x \circ s$ might not belong to K. For instance, let us consider an example for which $K \in \Re^3$ and $x = (2 + t, 1, 1)^T$, $s = (2 + t, 1, -1)^T \in K_+$, where t is some small positive number. Obviously one has $x \circ s = ((2 + t)^2, 4 + 2t, 0)^T \in K_+$. Moreover, it is easy to see $\lambda_{\min}(x \circ s) = 2t + t^2$. Thus, when t decreases to zero, the function $\Psi(x \circ s)$ goes to infinity. However, one can readily verify that for sufficiently small $t > 0$, both the functions $\Psi(x)$ and $\Psi(s)$ are bounded above. This examples shows that for $x, s \in K_+$, if $\det(x \circ s) \neq \det(x)\det(s)$, then the relation

$$\Psi(x \circ s) \leq \frac{1}{2}(\Psi(x^2) + \Psi(s^2))$$

might fail. We also mention that, in the SDO case, for any positive definite matrices X and S, since the matrix XS is diagonalizable and has positive eigenvalues, the function $\Psi(XS)$ (5.3) is well defined.

As we have observed in the LO and SDO cases, to establish the complexity of the algorithm, we need to bound the derivatives of certain proximity functions in suitable space. For SOCO, this requires us to discuss the derivatives of the functions $\psi(x(t))$ and $\Psi(x(t))$, where

$$x(t) = (x_1(t), ..., x_n(t))^T$$

is a mapping from $\Re$ into $\Re^n$. The rest of this section follows step by step the procedure in Section 5.2 for SDO. First, for simplicity, let us denote by $x'(t)$ the derivative of $x(t)$ with respect to t:

$$x'(t) = (x_1'(t), ..., x_n'(t))^T.$$

The following result provides the means to measure the first-order directional derivative of a general function $\Psi(x(t))$ and to bound its second derivative with respect to the variable t. Recall that we denote by $\psi'(x)$ the function given by (6.11) with a kernel function $\psi'(t)$.

Lemma 6.2.10 *Suppose that $x(t)$ is a mapping from $\mathfrak{R}$ into $\mathfrak{R}^n$. If $x(t)$ is twice differentiable with respect to t for all $t \in (l_t, u_t)$ and $\psi(t)$ is also a twice continuously differentiable function in a suitable domain that contains $\lambda_{\max}(x(t))$ and $\lambda_{\min}(x(t))$, then*

$$\frac{d}{dt}\mathrm{Tr}(\psi(x(t))) = \mathrm{Tr}(\psi'(x(t)) \circ x'(t)) \quad \forall t \in (l_t, u_t)$$

and

$$\frac{d^2}{dt^2}\mathrm{Tr}(\psi(x(t))) \leq \varpi \mathrm{Tr}(x'(t) \circ x'(t)) + \mathrm{Tr}(\psi'(x(t)) \circ x''(t)), \qquad (6.20)$$

where

$$\varpi = \max\{|\psi''(\lambda_{\max}(x(t)))|, |\psi''(\lambda_{\min}(x(t)))|, |\Delta\psi'(\lambda_{\max}(x(t)), \lambda_{\min}(x(t)))|\}.$$

Proof Without loss of generality we assume that $\|x_{2:n}\| > 0$. From definition (6.6) we conclude

$$\mathrm{Tr}(\psi(x(t))) = \psi(\lambda_{\max}(x(t))) + \psi(\lambda_{\min}(x(t))).$$

It follows that

$$\frac{d}{dt}\mathrm{Tr}(\psi(x(t))) = \psi'(\lambda_{\max}(x(t)))\left(x_1'(t) + \frac{1}{\|x_{2:n}(t)\|}x_{2:n}(t)^\mathrm{T}x_{2:n}'(t)\right)$$

$$+ \psi'(\lambda_{\min}(x(t)))\left(x_1'(t) - \frac{1}{\|x_{2:n}(t)\|}x_{2:n}(t)^\mathrm{T}x_{2:n}'(t)\right).$$

Now recalling definition (6.11), we obtain

$$\psi'(x(t))_1 := \frac{1}{2}(\psi'(\lambda_{\max}(x(t))) + \psi'(\lambda_{\min}(x(t)))),$$

$$\psi'(x(t))_{2:n} := \Delta\psi'(\lambda_{\max}(x(t)), \lambda_{\min}(x(t)))x_{2:n}.$$

By simple calculus, from the above two equalities one can readily check that

$$\frac{d}{dt}\mathrm{Tr}(\psi(x(t))) = 2\psi'(x(t))^\mathrm{T}x'(t) = \mathrm{Tr}(\psi'(x(t)) \circ x'(t)).$$

This proves the first statement of the lemma.

To prove the second statement of the lemma, we first observe that

$$\frac{d^2}{dt^2}\Psi(x(t)) = \mathrm{Tr}\left(\frac{d}{dt}\psi'(x(t)) \circ x'(t)\right) + \mathrm{Tr}(\psi'(t) \circ x''(t)).$$

It is straightforward to check that

$$\frac{d}{dt}\psi'(x(t)) = w_1 + w_2 + w_3,$$

where

$$w_1 = \frac{\psi''(\lambda_{\max}(x(t)))\left(x_1'(t) + \frac{x_{2:n}(t)^{\mathrm{T}}x_{2:n}'(t)}{\|x_{2:n}(t)\|}\right)}{2}\left(1, \frac{x_{2:n}(t)^{\mathrm{T}}}{\|x_{2:n}(t)\|}\right)^{\mathrm{T}},$$

$$w_2 = \frac{\psi''(\lambda_{\min}(x(t)))\left(x_1'(t) - \frac{x_{2:n}(t)^{\mathrm{T}}x_{2:n}'(t)}{\|x_{2:n}(t)\|}\right)}{2}\left(1, -\frac{x_{2:n}(t)^{\mathrm{T}}}{\|x_{2:n}(t)\|}\right)^{\mathrm{T}},$$

$$w_3 = \Delta\psi'(\lambda_{\max}(x(t)), \lambda_{\min}(x(t)))\left(0, x_{2:n}'(t)^{\mathrm{T}} - \frac{x_{2:n}(t)^{\mathrm{T}}x_{2:n}'(t)}{\|x_{2:n}(t)\|^2}x_{2:n}(t)^{\mathrm{T}}\right)^{\mathrm{T}}.$$

From the Cauchy–Schwarz inequality, we deduce that

$$\frac{x_{2:n}(t)^{\mathrm{T}}x_{2:n}'(t)}{\|x_{2:n}(t)\|} \le \|x_{2:n}'(t)\|.$$

This relation, together with the definition of ϖ, implies

$$\mathrm{Tr}((w_1 + w_2) \circ x'(t)) = \psi''(\lambda_{\max}(x(t)))\left(x_1'(t) + \frac{x_{2:n}(t)^{\mathrm{T}}x_{2:n}'(t)}{\|x_{2:n}(t)\|}\right)^2$$

$$+ \psi''(\lambda_{\min}(x(t)))\left(x_1'(t) - \frac{x_{2:n}(t)^{\mathrm{T}}x_{2:n}'(t)}{\|x_{2:n}(t)\|}\right)^2$$

$$\le 2\varpi\left((x_1'(t))^2 + \left(\frac{x_{2:n}(t)^{\mathrm{T}}x_{2:n}'(t)}{\|x_{2:n}(t)\|}\right)^2\right).$$

On the other hand, using simple calculus, one has

$$\mathrm{Tr}(w_3 \circ x'(t)) = 2\Delta\psi'(\lambda_{\max}(x(t)), \lambda_{\min}(x(t)))\left(\|x_{2:n}'(t)\|^2 - \left(\frac{x_{2:n}(t)^{\mathrm{T}}x_{2:n}'(t)}{\|x_{2:n}(t)\|}\right)^2\right)$$

$$\le 2\varpi\left(\|x_{2:n}'(t)\|^2 - \left(\frac{x_{2:n}(t)^{\mathrm{T}}x_{2:n}'(t)}{\|x_{2:n}(t)\|}\right)^2\right).$$

Adding the above two inequalities together, we obtain the desired relation (6.20). $\qquad\square$

It is worth considering the special case $K \in \Re^2$ where we can cast a SOCO problem as an SDO problem. Note that in the SDO case,

$$X(t) = \begin{pmatrix} x_1(t) & x_2(t) \\ x_2(t) & x_1(t) \end{pmatrix}.$$

In this situation, there holds trivially

$$\|X'(t)\|^2 = 2\|x'(t)\|^2 = \text{Tr}(x'(t) \circ x'(t)).$$

Recalling the difference between the definitions of $\Psi(x)$ by (6.15) and $\Psi(X)$ by (5.3), one can easily verify that the estimates given in Lemma 6.2.10 are precisely the same as those presented in its SDO analogue Lemma 5.2.8.

6.3 NEW SEARCH DIRECTIONS FOR SOCO

In the present section we consider various search directions used in IPMs for SOCO and introduce some new search directions based on self-regular proximities in the second-order cones.

6.3.1 Preliminaries

Most IPMs for SOCO employ different search directions together with suitable strategies to follow the central path appropriately. As in the SDO case, the search directions for SOCO are usually derived from certain Newton systems in various scaled spaces. To address this issue more clearly, we need to go into a little more detail. First we note that for the Jordan algebra $\circ$, x and s commute: $x \circ s = s \circ x$. This is different from the SDO situation, where the matrix pair X and S do not commute in general. From this fact, one might be encouraged to guess that the Newton system for SOCO is well defined if both $x \succ_K 0$ and $s \succ_K 0$ are feasible for SOCO. In the sequel we show that this conjecture might not be true.[5]

Consider the following linearized Newton system for (6.2):

$$\begin{pmatrix} A & 0 & 0 \\ 0 & E_n & A^T \\ \text{mat}(s) & \text{mat}(x) & 0 \end{pmatrix} \begin{pmatrix} \Delta x \\ \Delta s \\ \Delta y \end{pmatrix} = \begin{pmatrix} 0 \\ 0 \\ \mu\tilde{e} - \text{mat}(x)s \end{pmatrix}, \quad x, s \succ_K 0. \quad (6.21)$$

This system is well defined if and only if the Jacobian matrix within is

---

[5] One possible explanation for this phenomenon is that, although the Jordan algebra $\circ$ commutes, it does not satisfy the associative law. For example, let $x = (3, 1, 2)^T$, $s = (3, -1, -2)^T$ and $z = (1, 2, 1)^T$. One can easily check that $(x \circ s) \circ z \neq x \circ (s \circ z)$. On the other hand, the general multiplication operator for matrices is not commutative but associative, that is, for any $X, S, Z \in \Re^{n \times n}$, there holds $(XS)Z = X(SZ)$. We also mention that, as observed in the previous section of this chapter, for some $x, s \succ_{K0}$, the Jordan product $x \circ s$ might not be in K.

nonsingular. For simplicity, let us denote temporarily $X = \text{mat}(x)$ and $S = \text{mat}(s)$. We also denote by E_m and E_n the identity matrices in the suitable spaces. Now by using simple multiplication of matrices, one has

$$
\begin{pmatrix} A & 0 & 0 \\ 0 & E_n & A^\mathrm{T} \\ S & X & 0 \end{pmatrix} \begin{pmatrix} E_n & -S^{-1}X & 0 \\ 0 & E_n & 0 \\ 0 & 0 & E_m \end{pmatrix} \begin{pmatrix} E_n & 0 & 0 \\ 0 & E_n & -A^\mathrm{T} \\ 0 & 0 & E_m \end{pmatrix}
$$

$$
= \begin{pmatrix} A & -AS^{-1}X & AS^{-1}XA^\mathrm{T} \\ 0 & E_n & 0 \\ S & 0 & 0 \end{pmatrix}.
$$

This relation immediately gives the following result.

Lemma 6.3.1 *System (6.21) is well defined if and only if the matrix $AS^{-1}XA^\mathrm{T}$ is nonsingular.*

Recall that in the introduction of this chapter, we assume that the matrix A has full row rank. As a consequence, if the matrix $S^{-1}X$ is positive definite, then $AS^{-1}XA^\mathrm{T}$ is also positive definite and thus nonsingular. For LO, if the primal-dual pair x and s are both strictly feasible, then the matrix $S^{-1}X$ is diagonal and positive. Hence the Newton system for LO is well defined. In the case of SOCO, however, the matrix $AS^{-1}XA^\mathrm{T}$ may fail to be nonsingular. The reason is that if Z is a matrix with positive eigenvalues, then the symmetric matrix $Z + Z^\mathrm{T}$ may fail to be positive definite. We show this by an example. Consider the following SOCO problem with

$$
A = (0, \sqrt{3.69} + 0.7, 1), \quad K = \left\{ x \in \mathfrak{R}^3, x_1 \geq \sqrt{x_2^2 + x_3^2} \right\}.
$$

Assume we have a feasible primal-dual pair $(x := (1, 0.8, 0.5)^\mathrm{T}$, $s := (1, 0.7, 0.7))^\mathrm{T}$. Obviously $x, s \succ_K 0$. However, one can easily verify that $AS^{-1}XA^\mathrm{T} = 0$. This illustrates that system (6.21) is not well defined for this specific example.

6.3.2 Scaling Schemes for SOCO

As we observed in the preceding example, the linearized Newton system might not be well defined. To guarantee that the Newton-type system has a unique solution, people usually resort to some scaling schemes in the SDO case. In the sequel we introduce certain variants of such scaling schemes for SOCO. First it should be noted that in the rest of this section we consider the more general case $N > 1$. In this situation, the definitions $\psi(x)$, $\Psi(x)$ and the

Jordan algebra should be modified accordingly as follows:

$$\psi(x) = \left(\psi(x^1), \psi(x^2), ..., \psi(x^N)\right)^{\mathrm{T}}, \quad \Psi(x) = \sum_{i=1}^{N} \Psi(x^i), \qquad (6.22)$$

$$x \circ s = (x^1 \circ s^1, x^2 \circ x^2, ..., x^N \circ s^N)^{\mathrm{T}}. \qquad (6.23)$$

The scaling scheme for SOCO was first proposed and studied by Tsuchiya [113,114]. Before we discuss these scaling techniques, recall from section 6.2 that closely associated with each cone K^j are the matrices

$$E_{n_j} := \mathrm{diag}(1, 1, ..., 1), \quad Q^j = \mathrm{diag}(1, -1, ..., -1)$$

where E_{n_j} is the identity matrix in space $\Re^{n_j \times n_j}$ and Q^j is the representation matrix of the cone K^j, since

$$K^j = \left\{ x^j \in \Re^{n_j} : (x^j)^{\mathrm{T}} Q^j x^j \geq 0, \; x_1^j \geq 0 \right\}.$$

It is easy to see that $(Q^j)^2 = E_{n_j}$.

Now we are ready to give the definition of a scaling matrix for general second-order cones.

Definition 6.3.2   *A matrix $W^j \in \Re^{n_j \times n_j}$ is a scaling matrix for the cone K^j if it satisfies the condition*

$$W^j Q^j W^j = Q^j, \quad W^j \succ 0.$$

We remind the reader that here $W^j \succ 0$ means that W^j is positive definite and symmetric. In view of this definition, if W^j is a scaling matrix for the cone K^j, so is $(W_j)^{-1}$. In particular, for any $x \succ_K 0$, there is a scaling matrix W_x and diagonal matrix U_x such that $U_x W_x e = x$. Since the matrix mat(x) and W_x commute, one can easily see $U_x^{-1} W_x^{-1} e = x^{-1}$. For details, see [78,114].

A scaled pair $(\tilde{x}, \tilde{s})$ is obtained by the transformation

$$\tilde{x} = UWx, \quad \tilde{s} := (UW)^{-1}s,$$

where

$$W := \mathrm{diag}(W^1, ..., W^N), \quad U := \mathrm{diag}(u_1 E_{n_1}, ..., u_N E_{n_N}), \quad u_1, ..., u_N > 0.$$

Several elementary properties of such a transformation are summarized in the following proposition.

Proposition 6.3.3   *For any $j \in J$, we have*

(i) $\mathrm{Tr}(x^j \circ s^j) = \mathrm{Tr}(\tilde{x}^j \circ \tilde{s}^j)$;

(ii) $u_j^2 \det(x^j) = \det(\tilde{x}^j), \; u_j^{-2} \det(s^j) = \det(\tilde{s}^j)$;

(iii) $x \succ_K 0$ *(or $x \succeq_K 0$) if and only if $\tilde{x} \succ_K 0$ (or $\tilde{x} \succeq_K 0$).*

Proof The proof follows directly from the definition of scaling matrices. We refer to [8,113] for details. □

Let us define
$$\tilde{A} = A(UW)^{-1}, \quad \tilde{c} = (UW)^{-T}c.$$

We can rewrite system (6.22) in the scaled space as
$$\begin{cases} \tilde{A}\tilde{x} & = b, \\ \tilde{A}^{T}y + \tilde{s} & = \tilde{c}, \\ \mathrm{mat}(\tilde{x})\tilde{s} & = \mu\tilde{e}, \quad \tilde{x}, \tilde{s} \succeq_K 0. \end{cases} \tag{6.24}$$

If both x and s are strictly feasible for (SOCO) and (SOCD), so are the vectors $\tilde{x}$ and $\tilde{s}$ for the new SOCO problem in the scaled space. In this circumstance, the linearized Newton system for (6.24) amounts to solve the following system:
$$\begin{cases} \tilde{A}\tilde{d}_{x} & = 0, \\ \tilde{A}^{T}\tilde{d}_{y} + \tilde{d}_{s} & = 0, \\ \mathrm{mat}(\tilde{x})\tilde{d}_{s} + \mathrm{mat}(\tilde{s})\tilde{d}_{x} & = \mu\tilde{e} - \mathrm{mat}(\tilde{x})\tilde{s}, \quad \tilde{x}, \tilde{s} \succeq_K 0. \end{cases} \tag{6.25}$$

There are several popular choices for the scaling matrices W^{j} and the constants u_{j}. For instance, if UW is the identity matrix, then the solution of (6.25) yields the so-called AHO search direction [1]; if UW is chosen such that $\tilde{s} = \tilde{e}$ (or $\tilde{x} = \tilde{e}$), then we obtain the primal (or dual) HKM direction; if UW is chosen such that $\tilde{x} = \tilde{s}$, then we have the NT search direction [113,114]. Note that after the NT scaling, the matrix $\overline{A}\,\mathrm{mat}(\tilde{s})^{-1}\mathrm{mat}(\tilde{x})\overline{A}^{T} = \overline{A}\,\overline{A}^{T}$ is nonsingular. This further implies that system (6.25) is always well defined if both $x \succ_K 0$ and $s \succ_K 0$. Monteiro and Tsuchiya [78] have studied other search directions for SOCO.

6.3.3 Intermezzo: A Variational Principle for Scaling

As for the SDO case, we are particularly interested in the issue of which kind of scaling scheme is "optimal" under a certain variational principle. For this we consider the primal SOCO problem in the scaled space,

$$\text{(Scaled SOCO)} \qquad \min \mathrm{Tr}\!\left(\tilde{c}^{T}\tilde{x}\right)$$
$$\tilde{A}\tilde{x} = b, \quad \tilde{x} \succeq_K 0,$$

and its dual problem,

$$\text{(Scaled SOCD)} \quad \max b^{\mathrm{T}} y$$

$$\tilde{A}^{\mathrm{T}} y + \tilde{s} = \tilde{c}, \quad \tilde{s} \succeq_K 0.$$

We assume that a certain barrier method is employed to solve both the scaled primal and dual problems; namely, we minimize a specific potential function $\tilde{c}^{\mathrm{T}} \tilde{x} + \Psi(\tilde{x})$ and maximize $b^{\mathrm{T}} y - \Psi(\tilde{s})$, where $\Psi(\cdot)$ is a barrier function for the second-order cone K. The puzzle appearing here is for which kind of scaling matrix W and matrix U, the function

$$\tilde{c}^{\mathrm{T}} \tilde{x} - b^{\mathrm{T}} y + \Psi(\tilde{x}) + \Psi(\tilde{s})$$

attains its global minimal value. In the sequel we give a definite answer to this issue under the assumption that the corresponding kernel function $\psi(\cdot)$ satisfies condition SR2 strictly.

Proposition 6.3.4 *Suppose the functions $\psi(x)$ and $\Psi(\cdot)$ are defined by (6.11) and (6.22). If the function $\psi(t)$ satisfies condition SR2 strictly, that is, $\psi'(t) + t\psi''(t) > 0$ for any $t > 0$, then the function $\tilde{c}^{\mathrm{T}} \tilde{x} - b^{\mathrm{T}} y + \Psi(\tilde{x}) + \Psi(\tilde{s})$ attains its global minimal value with a scaling matrix W and a positive diagonal matrix U such that $\tilde{x} = \tilde{s}$.*

Proof First we observe that the inner product $\tilde{c}^{\mathrm{T}} \tilde{x}$ is invariant for any nonsingular matrices W and T. Thus we need only prove that $\Psi(\tilde{x}) + \Psi(\tilde{s})$ has a global minimizer when the matrices W and U are chosen so that $\tilde{x} = \tilde{s}$. We start with a discussion about the existence of such matrices W and U. For any $x, s \succ_K 0$ and $j \in J$, let us define

$$u_j := \left(\frac{\det(x^j)}{\det(s^j)} \right)^{1/4}, \tag{6.26}$$

$$w^j := \frac{u_j^{-1} s^j + u_j Q^j x^j}{\sqrt{2}\sqrt{\mathrm{Tr}(x^j \circ s^j) + \sqrt{\det(x^j)\det(s^j)}}}, \tag{6.27}$$

and

$$W_{NT}^j = \begin{pmatrix} w_1^j & (w_{2:n_j}^j)^{\mathrm{T}} \\ w_{2:n_j}^j & E_{n_j-1} + \frac{1}{1+w_1^j} w_{2:n_j}^j (w_{2:n_j}^j)^{\mathrm{T}} \end{pmatrix} = -Q^j + \frac{(\tilde{e}^j + w^j)(\tilde{e}^j + w^j)^{\mathrm{T}}}{1 + w_1^j}. \tag{6.28}$$

For the above choices, $\tilde{x}^j = u_j W_{NT}^j x^j = u_j^{-1} \left(W_{NT}^j \right)^{-1} s^j = \tilde{s}^j$ (see [8,113,114]). This proves the existence of such a scaling matrix W_{NT} and the matrix

$$U_{NT} = \text{diag}(u_1 E_{n_1}, \ldots, u_N E_{n_N}).$$

Hence it remains to show that for these specific choices of W_{NT} and U_{NT}, the value of the function $\Psi(\tilde{x}) + \Psi(\tilde{s})$ is optimal. To distinguish the NT scaling scheme from many other scaling schemes, we denote by

$$v = \begin{pmatrix} v^1 \\ \vdots \\ v^N \end{pmatrix} = \tilde{x}_{NT} = \tilde{s}_{NT} = \begin{pmatrix} u_1 W_{NT}^1 x^1 \\ \vdots \\ u_N W_{NT}^N x^N \end{pmatrix} = \begin{pmatrix} u_1^{-1}(W_{NT}^1)^{-1} s^1 \\ \vdots \\ u_N^{-1}(W_{NT}^N)^{-1} s^N \end{pmatrix}$$

the scaled vector based on the NT scaling, while $\tilde{x}$ and $\tilde{s}$ denote the scaled vectors using general scaling techniques. It follows that

$$\Psi(v) = \frac{1}{2} \sum_{j=1}^{N} \Big(\psi(\lambda_{\max}(v^j)) + \psi(\lambda_{\min}(v^j)) \Big).$$

Thus the proof will be finished if we can show that for any $j \in J$,

$$\psi(\lambda_{\max}(v^j)) + \psi(\lambda_{\min}(v^j)) \le \frac{1}{2}\Big(\psi(\lambda_{\max}(\tilde{x}^j)) + \psi(\lambda_{\min}(\tilde{x}^j)) \Big)$$

$$+ \frac{1}{2}\Big(\psi(\lambda_{\max}(\tilde{s}^j)) + \psi(\lambda_{\min}(\tilde{s}^j)) \Big) \qquad (6.29)$$

and the equality is true if and only if $\tilde{x}^j = \tilde{s}^j$. Now by recalling the definitions of the scaling matrices W_{NT} and U_{NT}, we can conclude that for any $j \in J$,

$$\text{Tr}\Big(x^j \circ s^j\Big) = \text{Tr}\Big(\tilde{x}^j \circ \tilde{s}^j\Big) = \text{Tr}\Big(v^j \circ v^j\Big) = \text{Tr}\Big([v^j]^2\Big), \qquad (6.30)$$

$$\det(v^j) = t_j^2 \det(x^j) = \sqrt{\det(x^j)\det(s^j)} = \sqrt{\det(\tilde{x}^j)\det(\tilde{s}^j)}. \qquad (6.31)$$

Thus, the vector v^j satisfies the requirements in the second statement of Proposition 6.2.9, where x and s are replaced by $\tilde{x}^j$ and $\tilde{s}^j$, respectively. Progressing in a similar vein as in the proof of the second statement of Proposition 6.2.9, we can obtain the desired relation (6.29), which concludes the proof of the proposition. $\Box$

For SOCO problems, a large-update IPM based on the NT search direction always has a theoretically lower iteration bound than large-update IPMs relying on other search directions [114].

6.3.4 New Proximities and Search Directions for SOCO

As a prelude to our new search direction, we next proceed to describe the NT-direction for SOCO. Assuming the current pair (x, s) is strictly feasible for both the primal problem (SOCO) and its dual (SOCD), the so-called NT direction for SOCO can be defined as follows. Let us denote

$$\overline{A} := \frac{1}{\sqrt{\mu}} A (U_{NT} W_{NT})^{-1}, \quad v := \frac{1}{\sqrt{\mu}} U_{NT} W_{NT} x,$$

$$d_x := \frac{1}{\sqrt{\mu}} U_{NT} W_{NT} \Delta x, \quad d_s := \frac{1}{\sqrt{\mu}} (U_{NT} W_{NT})^{-1} \Delta s.$$

Obviously $v \succ_K 0$. The NT search direction for SOCO is defined as the unique solution of the system

$$\begin{cases} \overline{A} d_x & = 0, \\ \overline{A}^T \Delta y + d_s & = 0, \\ d_x + d_s & = v^{-1} - v. \end{cases} \tag{6.32}$$

Before describing the new search direction for SOCO, we need to choose the proximity measure used in our new IPM for SOCO. Similar to the cases of LO and SDO, the proximity measure for SOCO is given by

$$\Phi(x, s, \mu) := \Psi(v) = \mathrm{Tr}(\psi(v)), \tag{6.33}$$

where $\psi(\cdot)$ is a univariate self-regular function.

The new search direction we propose for SOCO is a slight modification of the NT direction defined by the solution of the system

$$\begin{cases} \overline{A} d_x & = 0, \\ \overline{A}^T \Delta y + d_s & = 0, \\ d_x + d_s & = -\psi'(v). \end{cases} \tag{6.34}$$

Once we get d_x and d_s, we can compute Δx and Δs via (6.32). In view of the orthogonality of Δx and Δs, one can easily verify that

$$d_x^T d_s = 0. \tag{6.35}$$

We now discuss various properties of these self-regular proximities for SOCO. For simplicity, we need to introduce more notation. Let us denote by σ^2 the trace of the vector $\psi'(v)^2$, where $\psi'(v)$ is the vector on the right-hand side of the last equation of system (6.34). Therefore we have

$$\sigma^2 = \mathrm{Tr}(\psi'(v) \circ \psi'(v)) = 2\|\psi'(v)\|^2, \quad \sigma = \sqrt{2}\|\psi'(v)\|. \tag{6.36}$$

In the case of LO or SDO, the notation σ is defined as the norm of $\psi'(v)$ or $\psi'(V)$, separately. This is different from the SOCO situation. However, these definitions are consistent with respect to the definition of σ^2 given by the trace of the product $\mathrm{Tr}(\psi'(v)^2)$ for SOCO and $\mathrm{Tr}(\psi'(V)^2)$ for SDO. To facilitate the forthcoming analysis, we also define

$$\lambda_{\max}(v) = \max\{\lambda_{\max}(v^j) : j \in J\}, \quad \lambda_{\min}(v) = \min\{\lambda_{\min}(v^j) : j \in J\}.$$

Recall that by Lemma 6.2.6, we can deduce

$$\sigma^2 = \sum_{j=1}^{N}\left((\psi'(\lambda_{\max}(v^j)))^2 + (\psi'(\lambda_{\min}(v^j)))^2\right).$$

By using this relation and following similar steps to the proof of Proposition 3.1.5, we can prove the following results, which include several features of the proximity. These properties are naturally shared by general self-regular functions in the second-order cone K.

Proposition 6.3.5 *Let the proximity* $\Psi(v)$ *be defined by (6.33) and* σ *by (6.36). If the kernel function* $\psi(\cdot)$ *used in the proximity satisfies condition SR1 (see p. 29), then*

$$\Psi(v) \le \frac{\sigma^2}{2\nu_1}, \tag{6.38}$$

$$\lambda_{\min}(v) \ge \left(1 + \frac{q\sigma}{\nu_1}\right)^{-1/q}, \tag{6.39}$$

and

$$\lambda_{\max}(v) \le \left(1 + \frac{p\sigma}{\nu_1}\right)^{1/p}. \tag{6.40}$$

If $\lambda_{\max}(v) > 1$ *and* $\lambda_{\min}(v) < 1$, *then*

$$\sigma \ge \nu_1 \left(\frac{(\lambda_{\max}(v)^p - 1)^2}{p^2} + \frac{(\lambda_{\min}(v)^{-q} - 1)^2}{q^2}\right)^{1/2}. \tag{6.41}$$

For any $\vartheta > 1$,

$$\Psi(\vartheta v) \le \frac{\nu_2}{\nu_1}\left(\vartheta^{p+1}\Psi(v) + 2\vartheta\,\Upsilon'_{p,q}(\vartheta)\sqrt{N\nu_1\,\Psi(v)} + 2N\nu_1\Upsilon_{p,q}(\vartheta)\right). \tag{6.42}$$

If $\vartheta \in (1, 1 + \nu_3]$, *then*

$$\Psi(\vartheta v) \le \frac{\nu_2\nu_4}{\nu_1}\Psi(v) + \frac{2\nu_2\nu_4\sqrt{N\nu_1\,\Psi(v)}}{\nu_1}(\vartheta - 1) + 2N\nu_2\nu_4(\vartheta - 1)^2, \tag{6.43}$$

where ν_3, ν_4 *are the same constants as introduced in Corollary 2.2.6.*

We now discuss the relations between the duality gap and the proximity. By following a similar chain of reasoning as the proof of Lemma 2.2.3 we can easily deduce that

$$\frac{\Psi(v)}{\nu_1} \geq \frac{1}{2} \sum_{j=1}^{N} \left((\lambda_{\max}(v^j) - 1)^2 + (\lambda_{\min}(v^j) - 1)^2 \right)$$

$$= \|v\|^2 - \sum_{j=1}^{N} (\lambda_{\max}(v^j) + \lambda_{\min}(v^j)) + N \geq \|v\|^2 - 2\sqrt{N}\|v\| + N.$$

The above relation means that

$$\|v\| \leq \sqrt{N} + \sqrt{\frac{\Psi(v)}{\nu_1}}.$$

It readily follows that

$$\mathrm{Tr}(x \circ s) = 2\mu\|v\|^2 \leq 2N\mu + 4\mu\sqrt{\frac{N\Psi(v)}{\nu_1}} + \frac{2\Psi(v)}{\nu_1}\mu.$$

Therefore, $\mathrm{Tr}(x \circ s) = \mathcal{O}(N\mu)$ whenever $\Psi(v) = \mathcal{O}(N)$. In such a situation, the proximity plays the role of a potential function for minimizing the duality gap.

Let $x(\mu), s(\mu)$ be the targeted center in the primal and dual space. Obviously one has $s(\mu) = x(\mu)^{-1}/\mu$. Denote $W_{x(\mu)}$ the scaling matrix and $U_{x(\mu)}$ a diagonal matrix such that $U_{x(\mu)}W_{x(\mu)}e = x(\mu)$ and $\mu U_{x(\mu)}^{-1}W_{x(\mu)}^{-1}e = s(\mu)$. This indicates that $U_{s(\mu)} = \mu U_{x(\mu)}^{-1}$ and $W_{s(\mu)} = W_{x(\mu)}^{-1}$. We close this section by presenting a relation among values of the proximities $\Psi(v), \Psi(U_{x(\mu)}^{-1}W_{x(\mu)}^{-1}x)$ and $\Psi(U_{s(\mu)}^{-1}W_{s(\mu)}^{-1}s)$.

Lemma 6.3.6 *Suppose $x, s \succ_K 0$ and v is the scaled vector defined by the NT scaling. If the function $\psi(\cdot)$ used in (6.11) satisfies condition SR2 (2.2), then*

$$\Psi(v) \leq \frac{1}{2}\left(\Psi\left(U_{x(\mu)}^{-1}W_{x(\mu)}^{-1}x \right) + \Psi\left(U_{s(\mu)}^{-1}W_{s(\mu)}^{-1}s \right) \right).$$

Proof Denote $\tilde{x} = U_{x(\mu)}^{-1}W_{x(\mu)}^{-1}x$, $\tilde{s} = U_{s(\mu)}^{-1}W_{s(\mu)}^{-1}x$. One can readily show that

$$\mathrm{Tr}\left(v^2 \right) = \mathrm{Tr}\left(\tilde{x}^T \tilde{s} \right), \quad \det\left(v^2 \right) = \det\left(\tilde{x} \right) \det\left(\tilde{s} \right).$$

The remainder of the proof follows a similar procedure as the proof of Proposition 6.3.4 step by step. Thus the details are omitted here.   □

6.4 NEW POLYNOMIAL IPMS FOR SOCO

6.4.1 The Algorithm

The present section describes the new primal-dual algorithm for solving SOCO. By using the NT scaling, one can rewrite the centrality condition for SOCO as $v = \tilde{e}$. Consequently the neighborhood of the central path used in our new algorithm is also dependent on v. Denote

$$\mathcal{F}_{SOCO} = \{(x, s) \in K \times K : Ax = b, \ A^{\mathrm{T}}y + s = c\}.$$

We define the neighborhood of the central path as follows:

$$\mathcal{N}(\tau, \mu) = \{(x, s) : (x, s) \in \mathcal{F}_{SOCO}, \Phi(x, s, \mu) = \Psi(v) \le \tau\}.$$

Assuming that a starting point in a certain neighborhood of the central path is available, we can set out from this point. Actually, by using the so-called self-dual embedding model, one can further get the point exactly on the central path corresponding to $\mu = 1$ as an initial point (see [55,120] or Chapter 7). By solving system (6.34), one obtains a search direction. Then the iterate can be updated by means of a line search. If the current iterate goes beyond the neighborhood, then we utilize inner iterations to get a new iterate in the neighborhood. Otherwise we continue with the outer iteration and update μ by a fixed factor. The algorithm stops when the duality gap μ is sufficiently small and hence an approximate solution of the underlying problem is presented. The procedure of the new algorithm is outlined as follows.

Primal-Dual Algorithm for SOCO

Inputs
   A proximity parameter $\tau \ge \nu_1^{-1}$;
   an accuracy parameter $\varepsilon > 0$;
   a variable damping factor α;
   a fixed barrier update parameter $\theta \in (0, 1)$;
   (x^0, s^0) and $\mu^0 = 1$ such that $\Phi(x^0, s^0, \mu^0) \le \tau$.
begin
   $x := x^0$; $s := s^0$; $\mu := \mu^0$;
   while $N\mu \ge \varepsilon$ **do**
   begin
      $\mu := (1 - \theta)\mu$;
      while $\Phi(x, s, \mu) \ge \tau$ **do**
      begin
         Solve the system (6.34) for $\Delta x, \Delta y, \Delta s$;
         Determine a step size α;
         $x := x + \alpha \Delta x$;
         $y := y + \alpha \Delta y$;
         $s := s + \alpha \Delta s$;
      end
   end
end

Remark 6.4.1   *The algorithm will stop when an iterate satisfies $N\mu < \varepsilon$ and $\Phi(x, s, \mu) \leq \tau$. By recalling (6.44) we can claim*

$$x^{\mathrm{T}}s = \frac{1}{2}\,\mathrm{Tr}(x \circ s) \leq N\mu + 2\mu\sqrt{\frac{N\tau}{\nu_1}} + \mu\frac{\tau}{\nu_1}.$$

For instance, let us choose the parameter $\tau = N$ and the proximity satisfying condition SR1 with $\nu_1 = 1$. In such a case, the algorithm really works in a large neighborhood of the central path. One can easily verify that the algorithm will finally report a solution satisfying $x^{\mathrm{T}}s \leq 4\varepsilon$.

6.4.2 Complexity of the Algorithm

Having stated the algorithm in the previous section, we proceed to establish the polynomial complexity of the algorithm in the present section. As we already observed in early chapters for LO and SDO, a crucial step in the estimate of the algorithm's complexity is to evaluate how fast we can reduce the value of the proximity for a feasible step along the search direction.

Note that once the search direction $(\Delta x, \Delta s)$ is obtained, we need to decide how far we can go along this direction while staying in the feasible region. This amounts to estimating the maximal feasible step size. It should be noticed that for any step size α, the primal-dual pair $(x + \alpha\Delta x, s + \alpha\Delta s)$ is feasible if and only if the scaled primal-dual pair $(v + \alpha d_x, v + \alpha d_s)$ (see (iii) of Proposition 6.3.3) is feasible. In the sequel we give a certain sufficient condition for a step size to be strictly feasible and thus provide a lower bound for the maximal step size. To facilitate the analysis, for any $x^j \in \mathfrak{R}^{n_j}$, $j \in J$, we define

$$\lambda_{\max}(|x^j|) = |x_1^j| + \|x_{2:n_j}^j\|, \quad \lambda_{\min}(|x^j|) = |x_1^j| - \|x_{2:n_j}^j\|$$

and

$$\lambda_{\max}(|x|) = \max\{\lambda_{\max}(|x^j|) : j \in J\}, \quad \lambda_{\min}(|x|) = \min\{\lambda_{\min}(|x^j|) : j \in J\}.$$

A direct consequence of the above definitions is

$$\frac{1}{2}\sum_{j=1}^{N}\left(\lambda_{\max}(|x^j|)^2 + \lambda_{\min}(|x^j|)^2\right) = \|x\|^2, \quad x = (x^1, ..., x^N)^{\mathrm{T}}. \quad (6.45)$$

Now we have the following bounds.

Lemma 6.4.2   *Let $\alpha_{\max}$ be the maximal feasible step size and*

$$\bar{\alpha} = \lambda_{\min}(v)\sigma^{-1}. \quad (6.46)$$

Then

$$\alpha_{\max} \geq \overline{\alpha} \geq \sigma^{-1}\left(1 + \frac{q\sigma}{\nu_1}\right)^{-1/q}.$$

Proof First we observe that the primal-dual pair $(v + \alpha d_x, v + \alpha d_s)$ is strictly feasible if and only if for any $j \in J$,

$$v^j + \alpha d^j_x \succeq_{K^j} 0, \quad v^j + \alpha d^j_s \succeq_{K^j} 0.$$

For any fixed j, it is easy to see that

$$(v^j + \alpha d^j_x)_1 - \|(v^j + \alpha d^j_x)_{2:n_j}\| \geq v^j_1 - \|v^j_{2:n_j}\| - \alpha\lambda_{\max}(|d^j_x|)$$

$$\geq \lambda_{\min}(v) - \alpha\lambda_{\max}(|d_x|) \geq \lambda_{\min}(v) - \sqrt{2}\alpha\|d_x\|, \tag{6.47}$$

where the last inequality follows from (6.45). Now recalling the orthogonality relation (6.35), we can deduce

$$\|d_x\| \leq \|(d_x, d_s)\| = \sqrt{\|d_x\|^2 + \|d_s\|^2} = \frac{1}{\sqrt{2}}\sigma.$$

From inequality (6.39) in Proposition 6.3.5 we know that for any $\alpha \in [0, \overline{\alpha}]$,

$$\lambda_{\min}(v^j + \alpha d^j_x) \geq 0,$$

which is equivalent to $v^j + \alpha d^j_x \succeq_{K^j} 0$. Similarly one can show that for any $\alpha \in [0, \overline{\alpha}]$, there holds $v^j + \alpha d^j_s \succeq_{K^j} 0$. This completes the proof of the lemma.

$\square$

In view of Lemma 6.4.2, it is clear that we can use any $\alpha \in (0, \overline{\alpha})$ as a step size. Note that, after such a step, we get a new primal-dual pair $(x + \alpha\Delta x, s + \alpha\Delta s)$ or the scaled pair $(v + \alpha d_x, v + \alpha d_s)$ and then we need to use the NT scaling scheme to transform the primal and dual vectors to the same vector, which we denote by v^+. On the other hand, according to (6.33), the proximity after this step is defined as $\Psi(v^+)$. Let us denote the difference between the proximity before and after one step as a function of the step size, that is

$$g(\alpha) = \Psi(v^+) - \Psi(v). \tag{6.48}$$

The main task in the rest of this section is to study the decreasing behavior of $g(\alpha)$ for $\alpha \in [0, \overline{\alpha})$.

Since v^+ is the the scaled vector resulting from the NT scaling, from Proposition 6.3.3 we conclude that

$$\det\left(((v^+)^j)^2\right) = \det(v^j + \alpha d^j_x)\det(v^j + \alpha d^j_s), \quad j \in J,$$

and

$$\text{Tr}\left(((v^+)^j)^2\right) = \text{Tr}\left((v^j + \alpha d^j_x) \circ (v^j + \alpha d^j_s)\right), \quad j \in J.$$

Thus for any $j \in J$, the vectors $(v^+)^j, v^j + \alpha d_x^j$ and $v^j + \alpha d_s^j$ satisfy the requirement in the second statement of Proposition 6.2.9. Therefore, when the kernel function $\psi(\cdot)$ in (6.33) is self-regular, it follows readily from the second statement of Proposition 6.2.9 that

$$g(\alpha) = \Psi(v^+) - \Psi(v) \leq \frac{1}{2}(\Psi(v + \alpha d_x) + \Psi(v + \alpha d_s) - \Psi(v)) =: g_1(\alpha).$$

We proceed to estimate the decrease of the function $g_1(\alpha)$ for $\alpha \in [0, \bar{\alpha})$. For our specific purpose, we first estimate the first and second derivatives of $g_1(\alpha)$. From Lemma 6.2.10 it follows that

$$g_1'(\alpha) = \frac{1}{2}\operatorname{Tr}(\psi'(v + \alpha d_x) \circ d_x + \psi'(v + \alpha d_s) \circ d_s) \qquad (6.49)$$

and

$$g_1''(\alpha) = \frac{1}{2}\frac{d^2}{d\alpha^2}\operatorname{Tr}(\psi(v + \alpha d_x) + \psi(v + \alpha d_s)). \qquad (6.50)$$

The next result presents an upper bound for the second derivatives of $g_1'(\alpha)$. This result plays a crucial role in establishing the polynomial complexity of the algorithm.

Lemma 6.4.3 *Suppose that the kernel function $\psi(v)$ used in (6.33) is self-regular. Then for any $\alpha \in (0, \bar{\alpha})$,*

$$g_1''(\alpha) \leq \frac{1}{2}\nu_2\sigma^2\left((\lambda_{\max}(v) + \alpha\sigma)^{p-1} + (\lambda_{\min}(v) - \alpha\sigma)^{-q-1}\right).$$

Proof From Lemma 6.4.2 we know that the step size used in the lemma is strictly feasible. For any $j \in J$, let us denote $(v^+)_x^j = v^j + \alpha d_x^j$ and $(v^+)_s^j = v^j + \alpha d_s^j$. By means of Lemma 6.2.10 one gets

$$g_1''(\alpha) \leq \varpi_1 \|d_x\|^2 + \varpi_2 \|d_s\|^2,$$

where

$$\varpi_1 = \max_{i \in J}\left\{\left|\psi''\left((\lambda_{\max}(v^+)_x^j)\right)\right|, \left|\psi''\left(\lambda_{\min}((v^+)_x^j)\right)\right|, \left|\Delta\psi'\left(\lambda_{\max}((v^+)_x^j), (\lambda_{\max}((v^+)_x^j)\right)\right|\right\},$$

$$\varpi_2 = \max_{i \in J}\left\{\left|\psi''\left(\lambda_{\max}((v^+)_s^j)\right)\right|, \left|\psi''\left(\lambda_{\min}((v^+)_s^j)\right)\right|, \left|\Delta\psi'\left(\lambda_{\max}((v^+)_{s+}^j), (\lambda_{\max}((v^+)_s^j)\right)\right|\right\}.$$

Recall that $\sigma^2 = 2(\|d_x\|^2 + \|d_s\|^2)$. Thus the proof is done if we can show that

$$\max\{\varpi_1, \varpi_2\} \leq \nu_2(\lambda_{\max}(v) + \alpha\sigma)^{p-1} + (\lambda_{\min}(v) - \alpha\sigma)^{-q-1}. \qquad (6.51)$$

Combining the choice of ϖ_1 with the Mean Value Theorem [100], one can conclude that there exists a constant $\zeta_* \in [\lambda_{\min}(v), \lambda_{\max}(v)]$ such that

$$\varpi_1 = |\psi''(\zeta^*)|.$$

By using condition SR1 we immediately obtain

$$\varpi_1 \leq \nu_2(\zeta_*^{p-1} + \zeta_*^{-q-1}). \tag{6.52}$$

Now recalling the assumption $\alpha \in [0, \bar{\alpha}]$ and following a similar path as in the proof of (6.47), we can conclude that for any $j \in J$,

$$\lambda_{\min}(v^j) - \alpha\sigma \leq \lambda_{\min}(v^j) - \alpha\lambda_{\max}(|d_x|) \leq \lambda_{\min}(v^j + \alpha d_x^j)$$

and

$$\lambda_{\max}(v^j + \alpha d_x^j) \leq \lambda_{\max}(v^j) + \alpha\lambda_{\max}(|d_x|) \leq \lambda_{\max}(v) + \alpha\sigma.$$

It follows immediately that

$$\lambda_{\min}(v) - \alpha\sigma \leq \zeta_* \leq \lambda_{\max}(v) + \alpha\sigma,$$

which, together with (6.52), yields

$$\varpi_1 \leq \nu_2((\lambda_{\max}(v) + \alpha\sigma)^{p-1} + (\lambda_{\min}(v) - \alpha\sigma)^{-q-1}).$$

In an analogous vein one can deduce

$$\varpi_2 \leq \nu_2((\lambda_{\max}(v) + \alpha\sigma)^{p-1} + (\lambda_{\min}(v) - \alpha\sigma)^{-q-1}).$$

The above two inequalities give (6.51), which further gives the statement of the lemma.   □

The remaining discussions in this section are very similar to the LO and SDO cases. First we observe that by applying Lemma 6.2.10 to the function $g(\alpha)$, we readily claim

$$g'(0) = g_1'(0) = -\frac{\sigma^2}{2}.$$

From Lemma 6.4.3 it follows that

$g(\alpha) \leq g_1(\alpha)$

$$\leq -\frac{\sigma^2\alpha}{2} + \frac{1}{2}\nu_2\sigma^2 \int_0^\alpha \int_0^\xi ((\lambda_{\max}(v) + \zeta\sigma)^{p-1} + (\lambda_{\min}(v) - \zeta\sigma)^{-q-1})d\zeta d\xi,$$

which is essentially the same as its LO analogue (3.29), where the variables $v$max, $v$min are replaced by $\lambda_{\max}(v), \lambda_{\min}(v)$, respectively. Let us define

$$g_2(\alpha) := -\frac{\sigma^2\alpha}{2} + \frac{1}{2}\nu_2\sigma^2 \int_0^\alpha \int_0^\xi ((\lambda_{\max}(v) + \zeta\sigma)^{p-1} + (\lambda_{\min}(v) - \zeta\sigma)^{-q-1})d\zeta d\xi.$$

It is straightforward to verify that $g_2(\alpha)$ is strictly convex and twice differentiable for all $\alpha \in [0, \bar{\alpha})$. Let α^* be the unique global minimizer of $g_2(\alpha)$ in the interval $[0, \bar{\alpha})$, namely

$$\alpha^* = \arg\min_{\alpha \in [0,\bar{\alpha}]} g_2(\alpha), \tag{6.53}$$

or equivalently α^* is the unique solution of the equation

$$\frac{\nu_2}{p}\left((\lambda_{\max}(v) + \alpha^*\sigma)^p - \lambda_{\max}(v)^p\right) + \frac{\nu_2}{q}\left((\lambda_{\min}(v) - \alpha^*\sigma)^{-q} - \lambda_{\min}(v)^{-q}\right) = \sigma. \tag{6.54}$$

For this choice of α^*, by applying Lemma 1.3.3 we can readily claim that

$$g(\alpha^*) \le g_2(\alpha^*) \le \frac{1}{2}g_2'(0)\alpha^* = \frac{1}{2}g'(0)\alpha^*. \tag{6.55}$$

Thus it remains to estimate the value of α^*.

Lemma 6.4.4 *Let the constant α^* be defined by (6.53). Suppose that $\Psi(v) \ge \nu_1^{-1}$ and $v_{\max} > 1$ and let*

$$\nu_5 = \min\left\{\frac{\nu_1}{2\nu_2(p+\mu_1)+\nu_1(p-1)}, \frac{\nu_1^2}{(1+\nu_1)(2\nu_2(\nu_1+q)+\nu_1 q)}\right\}. \tag{6.56}$$

Then

$$\alpha^* \ge \nu_5 \sigma^{-(q+1)/q}. \tag{6.57}$$

In the special case where $\psi(t) = \Upsilon_{p,q}(t)$ is given by (2.5), the above bound simplifies to

$$\alpha^* \ge \min\left\{\frac{1}{3p+1}, \frac{1}{4+6q}\right\}\sigma^{-(q+1)/q}. \tag{6.58}$$

Proof See the proof of Lemma 3.3.3.   □

The following result estimates the decreasing value of the proximity when the step size α is given by α^* (6.53) or $\alpha = \nu_5 \sigma^{(q-1)/q}$. The proof of the theorem is similar to that for its LO analogue; thus the details are omitted here.

Theorem 6.4.5 *Let the function $g(\alpha)$ be defined by (6.48) with $\Psi(v) \ge \nu_1^{-1}$. Then the step size given by $\alpha = \alpha^*$ (6.53) or $\alpha = \nu_5 \sigma^{(q-1)/q}$ is strictly feasible. Moreover,*

$$g(\alpha) \le \frac{1}{2}g'(0)\alpha \le -\frac{\nu_5 \nu_1^{(q-1)/(2q)}}{4}\Psi(v)^{(q-1)/(2q)}.$$

In the special case where $\psi(t) = \Upsilon_{p,q}(t)$ is defined by (2.5) with $\nu_1 = \nu_2 = 1$, the above bound simplifies to

$$g(\alpha) \leq -\min\left\{\frac{1}{12p+4}, \frac{1}{24q+16}\right\}\Psi(v)^{(q-1)/(2q)}.$$

To get the total complexity result for the algorithm, we still need to describe the growth behavior of the proximity $\Psi(v)$. Suppose that the current point is in the neighborhood $\mathcal{N}(\mu, \tau)$, and we update μ to $(1-\theta)\mu$ for some $\theta \in (0,1)$. From its definition and proceeding as already done for the LO case, one can show that after the update of μ, the proximity is still bounded above by the number $\psi_0(\theta, \tau, 2N)$ defined by (3.40), where N is the number of cones. Immediately we have the following result.

Lemma 6.4.6 *Let $\Phi(x, s, \mu) \leq \tau$ and $\tau \geq \nu_1^{-1}$. Then after an update of the barrier parameter, no more than*

$$\left\lceil \frac{8q\nu_1^{-(q-1)/(2q)}}{\nu_5(q+1)}(\psi_0(\theta, \tau, 2N))^{(q+1)/(2q)} \right\rceil$$

iterations are needed to recenter. In the special case where $\psi(t) = \Upsilon_{p,q}(t)$ defined by (2.5) with $\nu_1 = \nu_2 = 1$, at most

$$\left\lceil \frac{8q\max\{3p+1, 6q+4\}}{q+1}(\psi_0(\theta, \tau, 2N))^{(q+1)/(2q)} \right\rceil$$

inner iterations are needed to recenter.

Thus the total complexity of the algorithm can be estimated as follows.

Theorem 6.4.7 *If $\tau \geq \nu_1^{-1}$, the total number of iterations required by the primal-dual Newton algorithm is not more than*

$$\left\lceil \frac{8q\nu_1^{-(q-1)/(2q)}}{\nu_5(q+1)}(\psi_0(\theta, \tau, 2N))^{(q+1)/(2q)} \right\rceil\left\lceil \frac{1}{\theta}\log\frac{N}{\varepsilon} \right\rceil.$$

In the special case where $\psi(t) = \Upsilon_{p,q}(t)$ is defined by (2.5) with $\nu_1 = \nu_2 = 1$, the total number of iterations required by the primal-dual Newton algorithm is less than or equal to

$$\left\lceil \frac{8q\max\{3p+1, 6q+4\}}{q+1}(\psi_0(\theta, \tau, 2N))^{(q+1)/(2q)} \right\rceil\left\lceil \frac{1}{\theta}\log\frac{N}{\varepsilon} \right\rceil.$$

Neglecting the influence of the constants in the expression in Theorem 6.4.7, one can safely conclude that for any fixed $\theta \in (0,1)$ with suitable $p, q \geq 1$, the algorithm with large-update for SOCO in the present section has an $\mathcal{O}(N^{(q+1)/(2q)}\log(N/\varepsilon))$ iterations bound, while the algorithm with small-update ($\theta = \mathcal{O}((1/\sqrt{N}))$) remains with the complexity of $\mathcal{O}(\sqrt{N}\log(N/\varepsilon))$

iterations bound. Furthermore, from Theorem 6.4.7 one can readily see that if p is a constant and $q = \log N$, then the new large-update algorithm has a complexity $\mathcal{O}(\sqrt{N} \log N \log(N/\varepsilon))$ iterations bound.

In closing this chapter we would like to point out that some constants (note that we replace n by $2N$ in the estimation of total inner iterations) appearing in the expression for the iterations bound of the SOCO algorithm are slightly larger than that for LO and SDO. One might wonder how this comes about. By way of illustration, let us consider a SOCO problem with only one two-dimensional second-order cone constraint, $K = \{(x_1, x_2)^T : x_1 \geq |x_2|\}$. Note that this constraint can be rewritten as linear inequalities by adding artificial variables as follows:

$$x_1 - x_2 - x_3 = 0, \quad x_1 + x_2 - x_4 = 0, \quad x_1, x_3, x_4 \geq 0.$$

Thus we finally have an LO problem with four variables and three of them are nonnegative while the original SOCO problem has only one conic constraint. In other words, the size of the reformulated LO problem is four times as large as that of the original SOCO problem. Taking into account this point, the larger constants in the estimate of the complexity of the algorithm for SOCO seem to be reasonable. On the other hand, as we mentioned in the introduction of this chapter, a SOCO problem can also be solved by casting it as a SDO problem in $S^{n \times n}$. In such a situation, the iteration complexity of the algorithm posed in this monograph for solving the reformulated SDO problem has a bound of $\mathcal{O}(n^{(q+1)/(2q)} \log(n/\varepsilon))$. If $K^j \in \Re^2$ for any $j \in J$, then one can see that the complexity of the algorithm for SOCO is precisely the same as that of its counterpart for SDO. However, it is straightforward to see that when $2N < n$, the algorithm that works directly on the original SOCO problem has a better iteration bound. The improvement is significant if $N \ll n$.

Chapter 7

Initialization: Embedding Models for Linear
Optimization, Complementarity Problems, Semidefinite
Optimization and Second-Order Conic Optimization

In this chapter, we describe diverse self-dual embedding models for various problems discussed in this monograph. These models provide us with powerful tools for locating a strictly feasible starting point as required by all feasible IPMs.

Like the algorithms proposed in the present work, many IPMs start with a strictly feasible initial point. However, obtaining such a point is usually as difficult as solving the underlying problem itself. On the other hand, it is desirable for a complete algorithm not only to identify the optimal solution when the problem is solvable, but also to detect infeasibility when the problem is either primal or dual infeasible. A robust and efficient way for handling these issues is to apply IPMs to an elegant augmented model: the so-called self-dual embedding model.

7.1 THE SELF-DUAL EMBEDDING MODEL FOR LO

Ye, Todd and Mizuno [128] introduced the homogeneous self-dual model for LO, which originated from the homogeneous model of Goldman and Tucker [32,115]. In [46] (see also [98]), Jansen, Roos and Terlaky proposed another slightly different but essentially equivalent self-dual model for LO. In the present section, for our specific algorithmic purpose, we only describe the model by Ye et al. All material in this section is extracted from [128] and the book [127]. Given any $x^0 > 0$, $s^0 > 0$ and y^0, with $\bar{n} := (x^0)^{\mathrm{T}}s^0 + 1$, the self-dual homogeneous model for LO can be written as follows:

$$
\begin{array}{llllllll}
(\mathrm{HMLO}) & & & \min & \bar{n}\vartheta & & & \\
& & Ax & -b\tau & +\bar{b}\vartheta & & & = 0, \\
& -A^{\mathrm{T}}y & & +c\tau & -\bar{c}\vartheta & -s & & = 0, \\
& b^{\mathrm{T}}y & -c^{\mathrm{T}}x & & +\gamma\vartheta & & -\rho & = 0, \\
& -\bar{b}^{\mathrm{T}}y & +\bar{c}^{\mathrm{T}}x & -\gamma\tau & & & & = -\bar{n}, \\
& y \in \Re^m, & x \geq 0, & \tau \geq 0, & \vartheta \in \Re, & s \geq 0, & \rho \geq 0, &
\end{array}
$$

where

$$\bar{b} = b - Ax^0, \quad \bar{c} = c - A^{\mathrm{T}}y^0 - s^0, \quad \gamma = c^{\mathrm{T}}x^0 + 1 - b^{\mathrm{T}}y^0. \quad (7.1)$$

Here $\bar{b}$, $\bar{c}$ and γ represent the "infeasibility" of the initial primal point, dual point and the duality gap, with respect to the original problems.

Note that by making use of the duality theory for general LO specialized to (HMLO), one can easily obtain its dual, which we denote by (HMLD), and one can easily see that (HMLO) has the same form as its dual (HMLD). In the sequel we present several conclusions about (HMLO). All these results are from [127,128]. For ease of reference we quote them here without proof.

Theorem 7.1.1 *Consider the problem (HMLO):*

(i) (HMLO) is self-dual, that is, (HMLO) is identical to its dual (HMLD).

(ii) *(HMLO) and thus (HMLD) has a strictly feasible point*

$$y = y^0, \quad x = x^0, \quad \tau = 1, \quad \vartheta = 1, \quad s = s^0, \quad \rho = 1.$$

(iii) *(HMLO) has an optimal solution and its optimal solution set is bounded.*

(iv) *The optimal value of (HMLO) is zero, and for any feasible $(y, x, \tau, \vartheta, s, \rho)$, the relation $x^{\mathrm{T}}s + \tau\rho = \bar{n}\vartheta$ holds.*

(v) *There is an optimal solution $(y^*, x^*, \tau^*, \vartheta^* = 0, s^*, \rho^*)$ such that $x^* + s^* > 0$ and $\tau^* + \rho^* > 0$. Such a solution is called a strictly self-complementary solution of (HMLO).*

Note that if we choose $x^0 = s^0 = e$ and $y^0 = 0$, then the resulting homogeneous LO problem has a strictly feasible solution on the central path with $\mu^0 = 1$. By Theorem 7.1.1, we can directly utilize our algorithm to solve (HMLO) and find an approximate solution to it. After that, an exact solution to (HMLO) can be found by applying a strongly polynomial rounding scheme as described in [98]. The following theorem demonstrates the relation between the optimal solutions of (HMLO) and those of (LP) (1.1) and (LD) (1.2).

Theorem 7.1.2 *Let $(y^*, x^*, \tau^*, \vartheta^* = 0, s^*, \rho^*)$ be a strictly self-complementary solution for (HMLO).*

(i) *(LP) (1.1) has an optimal solution if and only if $\tau^* > 0$. In this case, x^*/τ^* is an optimal solution for (LP) and $(y^*/\tau^*, s^*/\tau^*)$ is an optimal solution for (LD) (1.2).*

(ii) *(LP) has no optimal solution if and only if $\rho^* > 0$. In this situation, (x^*/ρ^*) or (s^*/ρ^*) or both are certificates for the infeasibility of (LP) or (LD). In particular if $c^{\mathrm{T}}x^* < 0$ then (LD) is infeasible; if $-b^{\mathrm{T}}y^* < 0$ then (LP) is infeasible; and if both $c^{\mathrm{T}}x^* < 0$ and $-b^{\mathrm{T}}y^* < 0$ hold, then both (LP) and (LD) are infeasible.*

Actually, one can further prove that any optimal solution $(y, x, \tau, \vartheta = 0, s, \rho)$ of (HMLO) with $\tau > 0$ yields an optimal solution for (LP) (1.1). If $\rho > 0$, then either (LP) or (LD) is infeasible.

Before closing this section, we remark that although we can easily get a strictly feasible starting point for the embedded problem, the computation of a Newton step for the new problem demands slightly more work than calculating a Newton step for the original problem (LP) (1.1). This is because we need to make two additional rank-1 updates. As a remedy for this point, in the practical implementation of standard IPMs for LO, a simplification of the above model has been employed to solve the LO problem [6,7]. This simplified model, which requires only one additional rank-1 update per iteration, was first proposed by Xu, Hung and Ye [123]. As observed by Roos, Terlaky and

Vial (see remark on page 422 of [98]), the simplified model is essentially equivalent to the original model. Because we can specify the value of ϑ to be the central path parameter μ and a full standard Newton step achieves the target duality gap, we are able to predict precisely the decreasing behavior of ϑ for a feasible step size via the change of the duality gap and thus remove variable ϑ completely from the system. This leads to a substantial improvement of the practical performance of the algorithm. We also note that if our new algorithm is employed to solve (HMLO), then the attractive connection between the variable ϑ and the central path parameter μ might fail to hold. Therefore, more work is needed to determine the role of the variable ϑ in our new algorithmic scheme and whether it can still be eliminated from our system.

7.2 THE EMBEDDING MODEL FOR CP

The homogeneous model for monotone CPs was first considered by Ye for LCPs [126] and later generalized to monotone NCPs by Andersen and Ye [9]. The model is a direct extension of its counterpart for LO. First let us recall that a mapping $f(x)$ is said to be monotone if and only if it is a $P_*(0)$ mapping. To proceed with our discussion, we need to introduce some useful concepts. As we mentioned earlier in chapter 1, there always exists a strictly complementary solution for LO, and the central path converges to such a solution. However, in the case of CPs, it is possible that the underlying CP has no strictly complementary solution. Instead of this, a so-called *maximally complementary solution* for monotone CPs does exist.

Definition 7.2.1 *A complementary solution* (x^*, s^*) *of CP is said to be a maximally complementary solution if the number of positive components in the vector* $x^* + s^*$ *is maximized over the solution set of the underlying CP.*

It has been proven [34,36] that any strictly feasible monotone CP must have at least one maximally complementary solution and that the central path of the CP converges to such a maximally complementary solution of the problem.

Bear in mind that IPMs are not omnipotent but are globally convergent only for some special classes of NCPs. An important class of NCPs that could be solved by IPMs in polynomial time is the class of CPs where the underlying mapping satisfies the so-called scaled Lipschitz condition [136].

Definition 7.2.1 *A monotone mapping* $f(x) : \mathfrak{R}^n_{++} \to \mathfrak{R}^n$ *satisfies the scaled Lipschitz condition if there is a nondecreasing function* $\nu(\alpha)$ *such that*

$$\|Xf(x + \Delta x) - f(x) - \nabla f(x)\Delta x\| \leq \nu(\alpha)\Delta x^\mathsf{T} \nabla f(x)\Delta x$$

for all $x \in \mathfrak{R}^n_{++}$ *and* Δx *satisfying* $\|X^{-1}\Delta x\| \leq \alpha$. *Here* $X = \mathrm{diag}(x)$.

Before introducing the embedding model for CPs, we consider first a new homogeneous mapping $\hat{f}(x)$:

$$\hat{f}(x, \tau) = \begin{pmatrix} \tau f(\frac{x}{\tau}) \\ -x^{\mathrm{T}} f(\frac{x}{\tau}) \end{pmatrix}, \quad \mathfrak{R}_+^n \times \mathfrak{R}_{++} \to \mathfrak{R}^{n+1}. \tag{7.2}$$

The following theorem presents several important features of the mapping $\hat{f}$ (see [9] for details).

Theorem 7.2.3 *Let $\hat{f}(x, \tau)$ be given by (7.2). Then*

(i) *$\hat{f}$ is a continuous homogeneous function in $\mathfrak{R}_+^n \times \mathfrak{R}_{++}$ with degree 1 and for any $(x; \tau) \in \mathfrak{R}_+^n \times \mathfrak{R}_{++}$*

$$(x; \tau)^{\mathrm{T}} \hat{f}(x, \tau) = 0$$

and

$$(x; \tau)^{\mathrm{T}} \nabla \hat{f}(x, \tau) = -\hat{f}(x, \tau).$$

(ii) *If $f(x)$ is a continuous monotone mapping from $\mathfrak{R}_+^n$ to $\mathfrak{R}^n$, then $\hat{f}(x, \tau)$ is a continuous monotone mapping from $\mathfrak{R}_+^n \times \mathfrak{R}_{++}$ to $\mathfrak{R}^{n+1}$.*

(iii) *If f satisfies the scaled Lipschitz condition, then so does $\hat{f}$.*

Now let us assume that $x^0 > 0$, $s^0 > 0$, $\tau^0 > 0$, $\rho^0 > 0$ are given and let

$$r^0 = s^0 - \tau^0 f\left(\frac{x^0}{\tau^0}\right), \quad z^0 = \rho^0 + (x^0)^{\mathrm{T}} f\left(\frac{x^0}{\tau^0}\right),$$

$$\bar{n} = (r^0)^{\mathrm{T}} x^0 + z^0 \tau^0 = (x^0)^{\mathrm{T}} s^0 + \tau^0 \rho^0.$$

Let us consider the following homogeneous augmented nonlinear CP:

(HMCP): $\quad \begin{pmatrix} s \\ \rho \\ \gamma \end{pmatrix} - \hat{f}(x, \tau, \vartheta) = \begin{pmatrix} s - \tau f(\frac{x}{\tau}) - \vartheta r^0 \\ \rho + x^{\mathrm{T}} f(\frac{x}{\tau}) - \vartheta z^0 \\ \gamma + (r^0)^{\mathrm{T}} x + z^0 \tau - \bar{n} - 1 \end{pmatrix} = 0,$

$$\begin{pmatrix} xs \\ \tau\rho \\ \vartheta\gamma \end{pmatrix} = 0, \quad \begin{pmatrix} x \\ \tau \\ \vartheta \end{pmatrix} \geq 0, \quad \begin{pmatrix} s \\ \rho \\ \gamma \end{pmatrix} \geq 0.$$

It is trivial to verify that the problem (HMCP) is a strictly feasible monotone CP. Further, by using result (ii) of Theorem 7.2.3 we can see that if f satisfies the scaled Lipschitz condition, then so does the augmented mapping $\hat{f}$. Note that we can further specify all the constants as $x^0 = s^0 = e$ and $\tau^0 = \rho^0 = \vartheta^0 = 1$. In this case, one gets an initial starting point on the central path with $\mu = 1$. Then we can utilize feasible IPMs to follow the central path and find a solution to (HMCP). It is also straightforward to check that for any solution point $(x^*, \tau^*, \vartheta^*, s^*, \rho^*, \gamma^*)$ of the problem (HMCP), equality $\vartheta^* = 0$ must hold. Further, if $\tau^* > 0$, then the point $(x^*/\tau^*, s^*/\tau^*)$ is a solution of the original CP. If $\rho^* > 0$, then the original CP is infeasible. Recall that the central path converges to a nontrivial maximally complementary solution of (HMCP). One can further show that if the original CP is solvable, then $\tau^* > 0$ must hold at a maximally complementary solution of (HMCP). Therefore, from a maximally complementary solution of (HMCP), we can further obtain a solution to the original CP when the underlying problem is solvable. However, when $\tau^* + \rho^* = 0$, then no infeasibility certificate exists. For more discussion about the relation between (HMCP) and the primitive CP, we refer readers to the paper [9] or the book [127]. In their paper, Andersen and Ye [9] considered a simplified version of the above model.

Unfortunately, it is not so easy to apply the above elegant model to $P_*(\kappa)$ CPs. To explain this point more clearly, we need to go into a little more detail. First we point out that, as stated in Theorem 7.2.3, when the mapping $f(x)$ is monotone, the augmented mapping $\hat{f}(x, \tau)$ is still monotone. However, there is no guarantee that the mapping $\hat{f}$ is again a $P_*(\kappa)$ mapping when $f(x)$ is a $P_*(\kappa)$ mapping. This can be verified via the following example [88]. Let

$$M = \begin{pmatrix} 0 & -4 & -1 \\ 1 & 0 & 0 \\ 8 & 0 & 0 \end{pmatrix}, \quad v = \begin{pmatrix} 1 \\ -2 \\ -4 \end{pmatrix}, \quad \hat{M} = \begin{pmatrix} M & v \\ -v^T & \gamma \end{pmatrix}, \quad \gamma > 0.$$

One can easily see that M is a $P_*(\kappa)$ matrix with $\kappa = 31/4$. However, the matrix $\hat{M}$ is not, because for $\hat{x} = (1/\gamma, 1, -1, 1/\gamma)^T \in \Re^4$, there hold

$$\hat{x}_i(\hat{M}\hat{x})_i \leq 0, \quad 1 \leq i \leq 4 \quad \text{and} \quad \hat{x}^T \hat{M} \hat{x} < 0.$$

Note that the matrix $\hat{M}$ has the same structure as the Jacobian matrix of the mapping $\hat{f}$. Thus from this example we can readily construct an affine $P_*(\kappa)$ mapping $f(x)$ so that the Jacobian matrix $\nabla \hat{f}$ of its augmented mapping $\hat{f}$ is not a $P_*(\kappa)$ matrix at some point in the suitable space. From Proposition 4.2.10 we can further conclude that $\hat{f}$ is not a $P_*(\kappa)$ mapping.

We remark that in their seminal work, Kojima et al. [58] proposed a big-$\mathcal{M}$ method to get a feasible starting point for an artificial LCP. However, their model relies obviously on the input data of the problem and thus their model can not be applied directly to the case of nonlinear $P_*(\kappa)$ CPs.

Slightly deviated from the aforementioned models, recently Peng, Terlaky and Yoshise [93a] proposed a new augmented model for general nonlinear CPs that can embed any CP into a strictly feasible CP in a higher dimensional space. This way we can apply IPMs directly to the augmented CP. From the solution of the embedded CP we can obtain a certificate for the solvability of the original CP. For details the reader is referred to [93a].

7.3 SELF-DUAL EMBEDDING MODELS FOR SDO AND SOCO

Several researchers have considered various embedding strategies for SDO and SOCO. For instance, Potra and Sheng [95] considered an extension of the simplified self-dual embedding model for LO to SDO; Luo, Sturm and Zhang [65] discussed the self-dual embedding model in the conic form, which includes SDO as a specific case; de Klerk, Roos and Terlaky [56] suggested adding some extra variables to get a strictly feasible starting point for the primal-dual SDO. In the sequel we first describe the self-dual embedding model in [56], which is a direct extension of its LO analogue (HMLO). We start with a basic definition.

Definition 7.3.1 *An optimal primal-dual pair (X^*, S^*) for an SDO problem is called maximally complementary if X^* and S^* have largest possible rank over the optimal solution set.*

As observed by many researchers [31,55,103], if the primal-dual pair of an SDO problem is strictly feasible, then there exist maximally complementary solutions and the central path of the underlying SDO problem converges to such a maximally complementary solution as the parameter μ goes to zero.

Now let us turn to the self-dual embedding scheme for SDO. For given $X^0 \succ 0$, $S^0 \succ 0$ and any $y^0 \in \Re^m$, we consider the following self-dual embedding SDO problem:

$$
\begin{aligned}
&\text{(HMSDO)} && \min && \bar{n}\vartheta && && \\
&&& \mathrm{Tr}(A_i X) \quad -b_i\tau \quad +\bar{b}_i\vartheta && && = 0, \quad i = 1,\dots,m, \\
&-\sum_{i=1}^m y_i A_i &&& +\tau C \quad -\vartheta\bar{C} \quad -S && = 0, \\
&b^{\mathrm{T}}y &&-\mathrm{Tr}(CX) && +\gamma\vartheta && -\rho = 0, \\
&-\bar{b}^{\mathrm{T}}y &&+\mathrm{Tr}(\bar{C}X) \quad -\gamma\tau && && = -\bar{n}, \\
&y \in \Re^m, && X \succeq 0, \quad \tau \geq 0, \quad \vartheta \in \Re, \quad S \succeq 0, \quad \rho \geq 0,
\end{aligned}
$$

where

$$
\bar{b}_i = b_i - \mathrm{Tr}\left(A_i X^0\right), \quad i = 1,\dots,m,
$$

$$\overline{C} = C - S^0 - \sum_{i=1}^{m} A_i y_i^0,$$

$$\gamma := 1 + \mathrm{Tr}\left(CS^0\right) - b^{\mathrm{T}} y^0,$$

$$\overline{n} = \mathrm{Tr}\left(X^0 S^0\right) + 1.$$

It is straightforward to see that a strictly feasible point for (HMSDO) is given by y^0, X^0, S^0 and $\vartheta^0 = \rho^0 = \tau^0 = \nu^0 = 1$. The following theorem from [56] summarizes several interesting properties of (HMSDO).

Theorem 7.3.2   *Consider the embedded problem (HMSDO). Then we have*

(i)   *(HMSDO) is self-dual.*

(ii)   *(HMSDO) is strictly feasible.*

(iii)   *(HMSDO) has a maximally complementary solution $(y^*, X^*, \tau^*, \vartheta^*, S^*, \rho^*, \nu^*)$.*

(iv)   *If $\tau^* > 0$, then X^*/τ^* is an optimal solution to the original SDO problem and $(y^*/\tau^*, S^*/\tau^*)$ is an optimal solution to its dual.*

(v)   *If $\rho^* > 0$, then either the original SDO problem or its dual is infeasible. Particularly, if $\mathrm{Tr}(CX^*) < 0$ then the dual (SDD) is infeasible, if $b^{\mathrm{T}} y^* > 0$ then (SDO) is infeasible.*

Thus, we can apply various feasible IPMs to locate an ε-approximate solution of (HMSDO), which provides us with an approximate solution to the original SDO problem or detects the infeasibility of either the primal or the dual problem. However, the nonlinearity of SDO gives rise to challenging difficulties to obtain an exact solution for SDO. Also, it might happen that at a maximally complementary solution of (HMSDO), $\tau^* = \rho^* = 0$. This further indicates that for the original SDO, there exists no complementarity solution pair and there is no certificate for the infeasibility of the considered SDO or its dual. This is different from the LO case but analogous to the case of nonlinear optimization. A detailed description of all these possible cases might involve a lot of quite complicated material that is irrelevant to our major topic in this work. Therefore, in lieu of tedious repetition, we simply omit such discussion and refer the interested reader to the papers [56,65] and the references therein.

Finally we describe briefly the self-dual embedding model for SOCO. For any point (x^0, s^0) satisfying $x^0, s^0 \in K_+$ and any $y^0 \in \Re^m$, let $\tau^0 = \rho^0 = 1$ and

$$\overline{b} = Ax^0 - b, \quad \overline{c} = A^{\mathrm{T}} y^0 + s^0 - c,$$

$$\gamma = -c^{\mathrm{T}} x^0 + b^{\mathrm{T}} y^0 - 1, \quad \overline{n} = (x^0)^{\mathrm{T}} s^0 + 1.$$

Then analogous to the SDO case we can obtain the following embedded SOCO problem:

$$
\begin{array}{lllllll}
\text{(HMSOCO)} & & \min & \bar{n}\vartheta & & & \\
& Ax & -b\tau & +\bar{b}\vartheta & & & = 0, \\
-A^\mathrm{T}y & & +c\tau & -\bar{c}\vartheta & -s & & = 0, \\
b^\mathrm{T}y & -c^\mathrm{T}x & & +\gamma\vartheta & & -\rho & = 0, \\
-\bar{b}^\mathrm{T}y & +\bar{c}^\mathrm{T}x & -\gamma\tau & & & & = -\bar{n}, \\
\end{array}
$$

$$y \in \Re^m, \qquad x \succeq_K 0, \quad \tau \geq 0, \quad \vartheta \in \Re, \quad s \succeq_K 0, \quad \rho \geq 0.$$

The results regarding (HMSOCO) are completely analogous to the (HMSDO) case as outlined by Theorem 7.3.2. In particular, one can easily see that the problem (HSOCO) is strictly feasible and self-dual. Hence, our new IPMs can be employed to find an approximate solution of (HSOCO), from which we can obtain either an approximate solution to the original SOCO problem, a certificate for the infeasibility of the original problem and its dual, or the evidence that neither an optimal solution pair with zero duality gap nor a certificate for unsolvability exists. The relations between a maximally complementary solution of (HSOCO) and the solution set of the original SOCO are analogous to the SDO case. For details, see [8,56,65].

Chapter 8

Conclusions

A summary of the main results of this work is presented here. Some sugges-tions for future research on IPMs are also addressed.

8.1 A SURVEY OF THE RESULTS AND FUTURE RESEARCH TOPICS

Starting from Karmarkar's remarkable paper [52], the study of IPMs has flourished in many areas of mathematical programming and greatly changed the state of the art in numerous areas of optimization theory and applications. As we mentioned in Chapter 1, Karmarkar's original scheme is closely associated with the classical logarithmic barrier approach to nonlinear problems. This vital observation led later to the development of the outstanding *self-concordancy* theory for classes of optimization problems [83]. Since the middle of the 1990s, the research on IPMs has shifted its focus from the theory of complexity to the application side of conic linear problems, with some unsolved puzzles left behind. One of these intriguing puzzles is the gap between the theoretical complexity and practical performance of small-update and large-update IPMs.

Aiming at introducing a new prototype for primal-dual IPMs and bridging the above-mentioned gap, we started Chapter 1 by describing various strategies used in general IPMs and giving some motivational observations. Then, in Chapter 2 we gave a thorough introduction to the theory of so-called univariate *self-regular* functions, paving the way for our novel approach of primal-dual IPMs in various spaces for solving diverse problems. The results in this chapter are collected mainly from the authors' works [90,91]. These include:

- A new class of functions is introduced and many fascinating features of these functions, such as the growth behavior, the barrier behavior, and the relations among the function itself and its derivatives, are explored. These interesting properties of self-regular functions enable us to design new large-update IPMs that enjoy better polynomial iteration bounds than the large-update IPMs designed and analyzed before.

- The new class of functions has a close relationship to the well-known class of *self-concordant* functions, but there are numerous striking differences as well.

These results of Chapter 2 present a new way to develop the theory of primal-dual IPMs beyond the celebrated theory of self-concordant functions. An interesting open issue is:

  o Are there other classes of functions suitable for the design of novel and efficient IPMs?

In [94], some new barrier functions beyond the set of self-regular functions are introduced and used in the design of new IPMs for LO. To be more specific, let us consider one of these functions:

$$\tilde{\psi}(t) = \frac{1}{2}(t^2 - 1) + \exp\left(\frac{1}{t} - 1\right) - 1.$$

One can readily check that this function is not *self-regular* but *self-concordant.*[1] Let us define a series of functions by

$$\psi_k(t) = \frac{1}{2}\left(t^2 - 1\right) + \left(1 + \frac{1}{k}\right)^{1-k} \cdot \left(\left(1 + \frac{1}{kt}\right)^k - \left(1 + \frac{1}{k}\right)^k\right), \quad k = 1, 2, \dots$$

By using Lemma 2.1.2 and simple calculus, we can show that the function $\psi_k(t)$ is *self-regular* for any $k \geq 1$. Furthermore, for any fixed $t > 0$, one has

$$\lim_{k \to \infty} \psi_k(t) = \tilde{\psi}(t).$$

This means, to some extent, that the function $\tilde{\psi}(t)$ can be approximated by a series of *self-regular* functions, or in other words, the function $\tilde{\psi}(t)$ is on the boundary of the convex cone of *self-regular* functions. As has been proved [94], large-update IPMs derived from those functions enjoy similar polynomial iteration bounds as those based on *self-regular* functions. Nevertheless, in the authors' view, more study is needed to explore various barrier functions that might give rise to new discoveries in the future study of IPMs.

Chapter 3 started with a straightforward extension of self-regular functions to the nonnegative orthant $\mathfrak{R}^n_+$. Then, a new class of proximities originating from *self-regular* functions, self-regular proximities for path-following methods for LO, was proposed and new search directions based on these proximities were suggested for solving the considered LO problem. By exploring the properties of *self-regular* functions and using a unified framework, we were able to prove the following conclusions:

- Large-update IPMs using our new directions enjoy polynomial complexity. The iteration bound of the algorithm is $\mathcal{O}(n^{(q+1)/(2q)} \log(n/\varepsilon))$, where $p, q \geq 1$ are constants, the so-called *growth degree* and *barrier degree* of the corresponding proximity, respectively, while small-update IPMs retain the best known $\mathcal{O}(\sqrt{n} \log(n/\varepsilon))$ iteration bound. In the very special case that p is a constant and $q = \log n$, the algorithm has an $\mathcal{O}(\sqrt{n} \log n \log(n/\varepsilon))$ iteration bound.

- Analogous complexity results can be derived for IPMs based on a more general class of proximities that requires only condition SR1.

It should be pointed out that in principle, our algorithms can be viewed as natural extensions of primal-dual potential reduction methods. A straightforward question arises:

- Is it possible to design pure primal (or dual) IPMs for LO based on some barrier functions similar to *self-regular* functions?

---

[1] One can further show that this function is not a *self-concordant barrier.*

Moreover, as observed in practice [7], software packages based on the so-called predictor-corrector IPMs are among the most efficient IPM solvers. Consider the issue:

o   How to incorporate some elegant ideas of predictor-corrector IPMs, such as suggested by Mehrotra [71], into our algorithmic design?

In this work we are biased towards the theoretical investigation of new IPMs. However, it is no less important to investigate the practical efficiency of these algorithms. Note that existing standard IPM solvers have so far worked very efficiently for most LO problems. For the time being, it is too early to claim that our new algorithms have better practical performance. However, we have implemented a simple version of our algorithm, and run a few problems. Our preliminary numerical results show that the algorithm is promising. Nevertheless, much more work is needed to test the new approach. We have observed both in our theoretical analysis and in our limited numerical experience, that the complexity of our algorithm depends heavily on the choice of the barrier function, or more concretely, relies on the *barrier degree q* and *growth degree p* of the proximity. For instance, when p is a constant and $q = \log(n)$, then our new large-update algorithm has an $\mathcal{O}(\sqrt{n}\log(n)\log(n/\varepsilon))$ iteration bound. As we observed in the introduction, there is a trade-off between the two basic objectives of general IPMs: tracing the central path and reducing the duality gap. The latter is the major target of the algorithm. In the present context, we place more emphasis on the first aspect, by using a higher-order barrier function. However, our preliminary numerical experience indicates that an algorithm with very large q does not work well. A practical issue is:

o   How to choose appropriate parameters (or combine the above two objectives together) for a given problem to achieve good practical performance?

Another interesting theoretical question is related to the proximity and the search direction. As we claimed in the introduction: "The proximity is crucial for both the quality and elegance of the analysis." An interesting observation mentioned at the end of section 3.4, is that even when the same search direction is employed, the proven complexity of the algorithm with large updates in this work is better than the one presented in [89]. In this work we focus mainly on new search directions and fail to improve the complexity of the standard large-update Newton method. Nevertheless, the above fact motivates us to ask:

o   Can we improve the complexity of the classical large-update Newton method by employing a new analysis based on some new proximity?

Although it is not so vitally relevant to the core of IPMs, the study of local convergence properties of IPMs has been an important topic in the IPM literature for a long time. This is particularly important because when the iterate is close to the solution set, the performance of an algorithm will be determined by its local convergence properties. Thus, it is worthwhile investigating:

○ Which kind of local convergence properties do our new IPMs have?

In Chapter 4 we discussed an extension of new IPMs based on the notion of *self-regularity* to large classes of CPs. The content of this chapter covers the material presented in [93]. The major conclusions include:

- A continuous differentiable mapping $f(x)$ is a $P_*(\kappa)$ mapping if and only if its Jacobian matrix $\nabla f(x)$ is a $P_*(\kappa)$ matrix for any $x \in \Re^n$.

- The new IPMs for solving smooth $P_*(\kappa)$ CPs have polynomial iteration bounds similar to their LO analogue.

The first conclusion gives a clear and affirmative answer to a fundamental issue concerning $P_*(\kappa)$ mappings. In the present monograph, we elaborate primarily on feasible IPMs. With regard to CPs, we note that in some specific cases, for instance in the case of monotone CPs, we can use the homogeneous model suggested by Andersen and Ye [9] to cast the original CP as a new augmented monotone CP for which a strictly feasible point can be readily obtained. If the involved mapping $f(x)$ is an affine $P_*(\kappa)$ mapping, then one could utilize the big-$\mathcal{M}$ method [58] to get a strictly feasible point for the formulated CP. However, it remains open how to find such a desired starting point for general nonlinear $P_*(\kappa)$ CPs. On the other hand, we also noticed that, to establish the polynomial complexity of the algorithm, we require that the mapping $f(x)$ satisfies some kind of smoothness condition. As proved by Andersen and Ye [9], a very interesting property of their model is that, if the original mapping $f(x)$ is monotone and satisfies the scaled Lipschitz condition, then so is the new mapping in the augmented homogeneous model. More research is needed to explore whether the new smoothness condition posed in Chapter 4 can be preserved when applied to Andersen and Ye's model.

Chapters 5 and 6 undertook the task of generalizing the theory of *self-regularity* and the *self-regularity* based IPMs from LO to LO over the second-order cone and over the cone of semidefinite matrices (SOCO and SDO). In principle, the results of these two chapters are taken from our works [91,92]. In summary:

- Fundamental properties about functions whose arguments are matrices or associated with the second-order cone have been established.

- The notions of self-regular functions and self-regular proximities have been extended to the conic cases and various features of these self-regular functions have been explored.

- New IPMs based on conic self-regular proximities have been introduced that enjoy the same polynomial iteration bound as their LO cousins.

Regarding the first point, these elementary conclusions about mappings associated with various cones are very interesting even from a pure mathematical point of view. They might be very helpful in the future study of some difficult optimization problems such as nonlinear optimization with matrix arguments. The proposed IPMs in these two chapters can be viewed as elegant combinations of *self-regularity* and the NT-scaling scheme (see Sections 5.3.2 and 6.3.4). A deep investigation of the proximities defined corresponding to various scaling schemes reveals that, among many other scaling schemes for SOCO and SDO, only the NT-scaling has an inherent connection with the notion of *self-regularity*. It remains an open question:

o   Can we design new efficient IPMs for SOCO and SDO based on other scaling techniques?

As in [90], the complexity of new IPMs can indeed be established without imposing the *self-regularity* requirement. It is also of interest to consider the issue:

o   Can we obtain similar complexity results for IPMs relying on some proximities whose kernel functions satisfy only Condition SR1?

Chapter 7 collects various self-dual embedding models for different problems considered in this context. These elegant models can be used to embed the underlying problem into a new problem so that we can directly apply feasible IPMs to the embedded problem and get an approximate solution to it. Moreover, from a solution of the embedded problem we can either extract an optimal solution of the original problem or get a certificate of the infeasibiltiy of the problem.

To summarize, the study of IPMs has flourished and matured during the period from the 1980s to the mid-1990s. Nevertheless, the results presented in this context convince us that the development of IPMs has not been exhausted yet. In this work, we have concentrated on the theoretical investigation of new IPMs. However, in view of the substantial flexibility in choosing the parameters in the algorithm and the use of the popular NT-scaling scheme, there is great potential to design practically efficient IPMs that are also capable of exploiting the special structure of the underlying problem. Extensive numerical tests and time should tell us whether our approach will highlight the future research on IPMs.

References

[1]   I. Adler and F. Alizadeh. Primal-dual interior-point algorithms for convex quadratically constrained and semidefinite optimization problems. Technical Report RRR-111-95, Rutcor, Rutgers Center for Operations Research, New Brunswick, NJ, 1995.

[2]   A. Alizadeh. Interior point methods in semidefinite programming with applications to combinatorial optimization. *SIAM Journal on Optimization*, 5:13–51, 1995.

[3]   F. Alizadeh, J.A. Haeberly, M.V. Nayakkankuppann, M. Overton and S. Schmieta. *SDPPack user's guide*. New York University, New York, 1997.

[4]   F. Alizadeh, J.P. Haeberly and M.L. Overton. A new primal-dual interior-point method for semidefinite programming, in: J.G. Lewis (editor), *Proceedings of the Fifth SIAM Conference on Applied Linear Algebra*, pp. 113–117, SIAM, 1994.

[5]   F. Alizadeh, J.-P.A. Haeberley and M.L. Overton. Primal-dual methods for semidefinite programming: convergence rates, stability and numerical results. *SIAM Journal on Optimization*, 8:746–768, 1998.

[6]   E.D. Andersen and K.D. Andersen. The MOSEK interior-point optimizer for linear programming: an implementation of the homogeneous algorithm, in: H. Frenk, C. Roos, T. Terlaky and S. Zhang (editors), *High Performance Optimization*, pp. 197–232, Kluwer Academic Publishers, Boston, MA, 1999.

[7]   E.D. Andersen, J. Gondzio, Cs. Mészáros and X. Xu. Implementation of interior-point methods for large scale linear programming, in: T. Terlaky (editor), *Interior Point Methods of Mathematical Programming*, pp. 189–252. Kluwer Academic Publishers, Dordrecht, The Netherlands, 1996.

[8]   E.D. Andersen, C. Roos and T. Terlaky. On implementing a primal-dual interior-point method for conic quadratic optimization. To appear in *Mathematical Programming*, 2002.

[9]   E.D. Andersen and Y. Ye. On a homogeneous algorithm for the monotone complementarity problem. *Mathematical Programming*, 84:375–399, 1999.

[10]   R. Bellman. *Introduction to Matrix Analysis, Volume 12. Classics in Applied Mathematics*, SIAM, Philadelphia, PA, 1995.

[11]   R. Bellman and K. Fan. On systems of linear inequalities in Hermitian matrix variables, in: V.L. Klee (editor), *Convexity*, 7:1–11, *Proceedings of Symposia in Pure Mathematics*, American Mathematical Society, Providence, RI, 1963.

[12]   A. Ben-Tal and A. Nemirovskii. Lectures on Modern Convex Optimization: Analysis, Algorithms and Engineering Applications, MPS-SAM Series on Optimisation, Vol. 02, SIAM, Philadelphia, PA, 2001.

[13]   K.H. Borgwardt. *The Simplex Method: A Probabilistic Analysis*. Springer-Verlag, Berlin, 1987.

[14]   W. Cook, A.M.H. Gerards, A. Schrijver and É. Tardos. Sensitivity results in integer linear programming. *Mathematical Programming, 34:251–264, 1986*.

[15]   R.W. Cottle, J.S. Pang, and R.E. Stone. *The Linear Complementarity Problem*. Academic Press Inc., San Diego, CA, 1992.

[16]   *CPLEX user's guide*, CPLEX Optimization Inc., Incline Village, NV, 1993.

[17] G.B. Dantzig. *Linear Programming and Extensions.* Princeton University Press, Princeton, NJ, 1963.

[18] I.I. Dikin. Iterative solution of problems of linear and quadratic programming. *Doklady Akademiia Nauk SSSK* 174, 747–748. Translated in *Soviet Mathematics Doklady*, 8, 674–675, 1967.

[19] J. Faraut and A. Korányi. *Analysis on Symmetric Cones.* Oxford University Press, New York, 1994.

[20] L. Faybusovich. Euclidean Jordan algebras and interior-point algorithms. *Positivity*, 1:331–357, 1997.

[21] L. Faybusovich. A Jordan-algebraic approach to potential-reduction algorithms. Technical Report, Department of Mathematics, University of Notre Dame, Notre Dame, IN, 1998.

[22] M.C. Ferris and J.S. Pang. Engineering and economic applications of complementarity problems, *SIAM Journal on Optimization*, 39:669–713, 1997.

[23] A.V. Fiacco and G.P. McCormick. *Nonlinear Programming: Sequential Unconstrained Minimization Techniques.* John Wiley & Sons, New York, 1968. Reprint: Volume 4 of *SIAM Classics in Applied Mathematics*, SIAM Publications, Philadelphia, PA, 1990.

[24] R. Frisch. The logarithmic potential method for solving linear programming problems. Memorandum, University Institute of Economics, Oslo, 1955.

[25] R. Frisch. The logarithmic potential method for convex programming problems. Memorandum, University Institute of Economics, Oslo, 1955.

[26] K. Fujisawa, M. Kojima and K. Nakata. *SDPA* (semidefinite programming algorithm) user's manual-version 4.10, Technical Report, Department of Information Science, Tokyo Institute of Technology, Tokyo, Japan 1998.

[27] M. Fukushima, Z.Q. Luo and P. Tseng. Smoothing functions for second-order-cone complementarity problems. Preprint, Department of Mathematics, University of Washington, Seattle, WA, November, 2000.

[28] P.E. Gill, W. Murray, M.A. Saunders, J.A. Tomlin and M.H. Wright. On projected Newton barrier methods for linear programming and an equivalence to Karmarkar's projective method. *Mathematical Programming*, 36:183–209, 1986.

[29] F. Glineur. Improving complexity of structured convex optimization problems using self-concordant barriers. To appear in European Journal of Operations Research, 2002.

[30] M.X. Goemans and D.P. Williamson. Improved approximation algorithms for maximum cut and satisfiability problems using semidefinite programming. *JACM*, 42:1115–1145, 1995.

[31] D. Goldfarb and K. Scheinberg. Interior point trajectories in semidefinite programming. *SIAM Journal on Optimization*, 8:871–886, 1998.

[32] A.J. Goldman and A.W. Tucker. Theory of linear programming, in: H.W. Kuhn and A.W. Tucker (editors), *Linear Inequalities and Related Systems, Annals of Mathematical Studies*, 38:63–97, Princeton University Press, Princeton, NJ, 1956.

[33] C.C. Gonzaga. Path-following methods for linear programming. *SIAM Review*, 34:167–227, 1992.

[34] O. Güler. Existence of interior-points and interior paths in nonlinear monotone complementarity problems. *Mathematics of Operations Research*, 18:128–147, 1993.

[35] O. Güler. Limiting behavior of the weighted central paths in linear programming. *Mathematical Programming*, 65:347–363, 1994.

[36] O. Güler and Y. Ye. Convergence behavior of interior-point algorithms. *Mathematical Programming*, 60:215–228, 1993.

[37] P.T. Harker and J.S. Pang. Finite-dimensional variational inequality and nonlinear complementarity problems: a survey of theory, algorithms and applications, *Mathematical Programming*, 48:161–220, 1990.

[38] C. Helmberg, F. Rendl, R.J. Vanderdei and H. Wolkowicz. An interior-point method for semidefinite programming. *SIAM Journal on Optimization*, 6:342–361, 1996.

[39] D. den Hertog. *Interior Point Approach to Linear, Quadratic and Convex Programming, Volume 277, Mathematics and its Applications*. Kluwer Academic Publishers, Dordrecht, The Netherlands, 1994.

[40] R.A. Horn and C.R. Johnson. *Topics in Matrix Analysis*. Cambridge University Press, 1991.

[41] P. Huard. Resolution of mathematical programming with nonlinear constraints by the method of centers, in: J. Abadie (editor), *Nonlinear Programming*, North-Holland Publishing Company, Amsterdam, The Netherlands, pp. 207–219, 1967.

[42] P. Hung and Y. Ye. An asymptotically $\mathcal{O}(\sqrt{n}L)$-iteration path-following linear programming algorithm that uses long steps. *SIAM Journal on Optimization*, 6:570–586, 1996.

[43] T. Illés, C. Roos and T. Terlaky. Polynomial affine-scaling algorithms for $P_*(\kappa)$ linear complementarity problems, in: P. Gritzmann, R. Horst, E. Sachs and R. Tichatschke (editors), *Recent Advances in Optimization*, Proceedings of the 8th French-German Conference on Optimization, Trier, July 21–26, 1996, Lecture Notes in Economics and Mathematical Systems 452, pp. 119–137, Springer-Verlag, 1997.

[44] G. Isac. *Complementarity Problems, Lecture Notes in Mathematics*, Springer-Verlag, New York, 1992.

[45] B. Jansen. *Interior Point Techniques in Optimization, Complexity, Sensitivity and Algorithms*. Kluwer Academic Publishers, Dordrecht, 1997.

[46] B. Jansen, C. Roos and T. Terlaky. The theory of linear programming: skew symmetric self-dual problems and the central path. *Optimization*, 29:225–233, 1994.

[47] B. Jansen, C. Roos, T. Terlaky and J.-Ph. Vial. Primal-dual algorithms for linear programming based on the logarithmic barrier method. *Journal of Optimization Theory Application*, 83:1–26, 1994.

[48] B. Jansen, C. Roos, T. Terlaky and Y. Ye. Improved complexity using higher order correctors for primal-dual Dikin affine scaling. *Mathematical Programming, Series B*, 76:117–130, 1997.

[49] B. Jansen, C. Roos, T. Terlaky and A. Yoshise. Polynomiality of primal-dual affine scaling algorithms for nonlinear complementarity problems. *Mathematical Programming*, 78:315–345, 1997.

[50] F. Jarre. The Method of Analytic Centers for Smooth Convex Programs. Ph.D Thesis, Institut für Angewandte Mathematik and Statistik, Universität Würzburg, Würzburg, Germany, 1989.

[51] F. Jarre. Interior-point methods via self-concordance or relative Lipschitz condition. *Optimization Methods Software*, 5:75–104, 1995.

[52] N.K. Karmarkar. A new polynomial-time algorithm for linear programming. *Combinatorica*, 4:373–395, 1984.

[53] L.G. Khachiyan. A polynomial algorithm for linear programming. *Doklady Akad. Nauk USSR*, 244:1093–1096, 1979. Translated in *Soviet Math. Doklady*, 20:191–194, 1979.

[54] V. Klee and G.J. Minty. How good is the simplex method? in: O. Shisha (editor), *Inequalities III*, Academic Press, New York, 1972.

[55] E. de Klerk. Interior Point Methods for Semidefinite Programming. Ph.D. Thesis, Faculty of ITS/TWI, Delft University of Technology, The Netherlands, 1997.

[56] E. de Klerk, C. Roos and T. Terlaky. Infeasible-start semidefinite programming algorithms via self-dual embeddings, in: P. Pardalos and H. Wolkowicz (editors), *Topics in Semidefinite and Interior-point Methods*, Toronto, ON, 1996, pp. 215–236, Fields Inst. Commun., 18, American Mathematical Society, Providence, RI, 1998.

[57] M. Kojima, N. Megiddo and T. Noma. Homotopy continuation methods for nonlinear complementarity problems. *Mathematics of Operations Research*, 16:754–774, 1991.

[58] M. Kojima, N. Megiddo, T. Noma and A. Yoshise. *A Unified Approach to Interior Point Algorithms for Linear Complementarity Problems*, Volume 538, Lecture Notes in Computer Science. Springer Verlag, Berlin, 1991.

[59] M. Kojima, S. Mizuno and A. Yoshise. A primal-dual interior-point algorithm for linear programming, in: N. Megiddo (editor), *Progress in Mathematical Programming: Interior Point and Related Methods*, pp. 29–47. Springer Verlag, New York, 1989.

[60] M. Kojima, S. Mizuno and A. Yoshise. A polynomial-time algorithm for a class of linear complementarity problems. *Mathematical Programming*, 44:1–26, 1989.

[61] M. Kojima, S. Shindoh and S. Hara. Interior-point methods for the monotone semidefinite linear complementarity problem in symmetric matrices. *SIAM Journal on Optimization*, 7:86–125, 1997.

[62] G. Lesaja. Interior-Point Methods for P_*-complementarity Problems. Ph.D. Thesis, University of Iowa, Iowa City, IA, 1996.

[63] M.S. Lobo, L. Vandenberghe, S. Boyd and H. Lebret. Applications of second-order cone programming. *Linear Algebra and its Applications*, 284:193–228, 1998.

[64] F.A. Lootsma. *Numerical Methods for Nonlinear Optimization*. Academic Press, London, 1972.

[65] Z.Q. Lou, J. Sturm and S. Zhang. Conic linear programming and self-dual embedding. *Optimization Methods and Software*, 14:169–218, 2000.

[66] I.J. Lustig, R.E. Marsten and D.F. Shanno. Interior point methods: computational state of the art. *ORSA Journal on Computing*, 6:1–15, 1994.

[67] O.L. Mangasarian. *Nonlinear Programming*, McGraw-Hill, New York, 1969. Reprinted by SIAM Publications, 1995.

[68] L. McLinden. The analogue of Moreau's proximation theorem, with applications to the nonlinear complementarity problems. *Pacific Journal of Mathematics*, 88:101–161, 1980.

[69] K.A. McShane, C.L. Monma and D.F. Shanno. An implementation of a primal-dual interior-point method for linear programming. *ORSA Journal on Computing*, 1:70–83, 1989.

[70] N. Megiddo. Pathways to the optimal set in linear programming, in: N. Megiddo (editor), *Progress in Mathematical Programming: Interior Point and Related Methods*, pp. 131–158. Springer Verlag, New York, 1989. Identical version in *Proceedings of the 6th Mathematical Programming Symposium of Japan*, Nagoya, Japan, pp. 1–35, 1986.

[71] S. Mehrotra. On the implementation of a (primal-dual) interior-point method. *SIAM Journal on Optimization*, 2:575–601, 1992.

[72] S. Mehrotra and Y. Ye. On finding the optimal facet of linear programs. *Mathematical Programming*, 62:497–515, 1993.

[72a] S. Mehrota and J. Sun. An algorithm for convex quadratic programming that requires $\theta(n^{3.5}L)$ arithmetic operations. *Mathematics of Operations Research*, 15:342–363, 1990.

[72b] S. Mehrota and J. Sun. A method of analytic centers for quadratically constrained convex quadratic programs, *SIAM Journal of Numerical Analysis*, 28:529–544, 1991.

[73] S. Mizuno and A. Nagasawa. A primal-dual affine scaling potential reduction algorithm for linear programming. *Mathematical Programming*, 62:119–131, 1993.

[74] R.D.C. Monteiro. Primal-dual path-following algorithms for semidefinite programming. *SIAM Journal on Optimization*, 7:663–678, 1997.

[75] R.D.C. Monteiro and I. Adler. Interior-path following primal-dual algorithms: Part I: Linear programming. *Mathematical Programming*, 44:27–41, 1989.

[76] R.D.C. Monteiro and I. Adler. Interior-path following primal-dual algorithms: Part II: Convex quadratic programming. *Mathematical Programming*, 44:43–66, 1989.

[77] R.D.C. Monteiro, I. Adler and M.G.C. Resende. A polynomial-time primal-dual affine scaling algorithm for linear and convex quadratic programming and its power series. *Mathematics of Operations Research*, 15:191–214, 1990.

[78] R.D.C. Monteiro and T. Tsuchiya. Polynomial convergence of primal-dual algorithms for the second-order cone program based on the MZ-family of directions. *Mathematical Programming*, 88:61–83, 2000.

[79] J. Moré and W. Rheinboldt. On *P*- and *S*-functions and related classes of *n*-dimensional nonlinear mappings. *Linear Algebra and its Applications*, 6:45–68, 1973.

[80] K.G. Murty. *Linear Programming*. John Wiley & Sons, Inc., 1983.

[81] A.S. Nemirovskii and K. Scheinberg. Extension of Karmarkar's algorithm onto convex quadratically constrained quadratic programming. *Mathematical Programming*, 72:273–289, 1996.

[82] Y.E. Nesterov. Quality of semidefinite relaxation for nonconvex quadratic optimization. *Optimization Methods and Software*, 9:141–160, 1998.

[83] Y.E. Nesterov and A.S. Nemirovskii. Interior point polynomial algorithms in convex programming. *SIAM Studies in Applied Mathematics*, Volume 13, SIAM, Philadelphia, PA, 1994.

[84] Y.E. Nesterov and M.J. Todd. Self-scaled barriers and interior-point methods for convex programming. *Mathematics of Operations Research*, 22:1–42, 1997.

[85] Y.E. Nesterov and M.J. Todd. Primal-dual interior-point methods for self-scaled cones. *SIAM Journal on Optimization*, 8:324–362, 1998.

[86] *Optimization Subroutine Library, Guide and References*. IBM Corporation, Kingston, NY, 1991.

[87] J.S. Pang. Complementarity problems, in: R. Horst and P. Pardalos (editors), *Handbook in Global Optimization*, pp. 271–338, Kluwer Academic Publishers, Boston, MA, 1994.

[88] J. Peng, C. Roos and T. Terlaky. New complexity analysis of primal-dual Newton methods for $P_*(\kappa)$ linear complementarity problems, in: H. Frenk, C. Roos, T. Terlaky and S. Zhang (editors), *High Performance Optimization*, pp. 245–265, Kluwer Academic Publishers, Boston, MA, 1999.

[89] J. Peng, C. Roos and T. Terlaky. New complexity analysis of the primal-dual Newton method for linear programming. *Annals of Operations Research*, 99:23-39, 2000.

[90] J. Peng, C. Roos and T. Terlaky. A new class of polynomial primal-dual methods for linear and semidefinite optimization. To appear in *European Journal of Operations Research*, 2002.

[91] J. Peng, C. Roos and T. Terlaky. Self-regular proximities and new search directions for linear and semidefinite optimization. To appear in *Mathematical Programming*, 2002.

[92] J. Peng, C. Roos and T. Terlaky. New primal-dual algorithms for second-order conic optimizaion based on self-regular proximities. To appear in *SIAM Journal on Optimization*, 2002.

[93] J. Peng, C. Roos, T. Terlaky and A. Yoshise. Self-regular proximities and new search directions for $P_*(\kappa)$ complementarity problems. AdvOL-Report#2000-6, Department of Computing and Software, McMaster University, Canada.

[93a] J. Peng, T. Terlaky and A. Yoshise. Some results about $P_*(k)$ complementarity problems. AdvOL-Report#2002-1, Department of Computing and Software, McMaster University, Canada.

[94] C. Roos. A comparative study of barrier functions for primal-dual interior-point algorithms in linear optimization. Technical Report in preparation, Faculty of Technical Mathematics and Informatics, Delft University of Technology, The Netherlands, 2001.

[95] F. Potra and R.Q. Sheng. Homogeneous interior-point algorithms for semidefinite programming. *Optimization Methods and Software*, 9:161–184, 1998.

[96] J. Renegar. A polynomial-time algorithm, based on Newton's method, for linear programming. *Mathematical Programming*, 40:59–93, 1988.

[97] J. Renegar. A Mathematical View of Interior-Point Methods in Convex Optimization. MPS-SIAM Series on Optimization, vol. 03, SIAM, Philadelphia, PA, 2001.

[98] C. Roos, T. Terlaky and J.-Ph.Vial. *Theory and Algorithms for Linear Optimization. An Interior Approach*. John Wiley & Sons, Chichester, 1997.

[99] C. Roos and J.-Ph.Vial. A polynomial method of approximate centers for linear programming. *Mathematical Programming*, 54:295–305, 1992.

[100] W. Rudin, *Principles of Mathematical Analysis*. Mac-Graw Hill Company, New York, 1978.

[101] M. Shida, S. Shindoh and M. Kojima, Existence of search directions in interior-point algorithms for the SDP and the monotone SDLCP, *SIAM Journal on Optimization*, 8:387–396, 1998.

[102] G. Sonnevend. An "analytic center" for polyhedrons and new classes of global algorithms for linear (smooth, convex) programming, in: A. Prèkopa, J. Szelezsán and B. Strazicky (editors), *System Modelling and Optimization: Proceedings of the 12th IFIP Conference*, Budapest, Hungary, September 1985, Volume 84, Lecture Notes in Control and Information Sciences, pp. 866–876. Springer Verlag, Berlin, 1986.

[103] J.F. Sturm. Theory and Algorithms of Semidefinite Programming, in: H. Frenk, C. Roos, T. Terlaky and S. Zhang (editors), *High Performance Optimization*, pp. 1–194, Kluwer Academic Publishers, Boston, MA, 1999.

[104] J.F. Sturm. Using SeDuMi 1.02, A MATLAB toolbox for optimization over symmetric cones. Interior point methods. *Optimization Methods and Software*, 11/12:625–653, 1999.

[105] J.F. Sturm and S. Zhang, Symmetric primal-dual path following algorithms for semidefinite programming, *Applied Numerical Mathematics*, 29:301–315, 1999.

[106] K. Tanabe. Centered Newton methods for mathematical programming, *System Modelling and Optimization*, 197–206, Springer-Verlag, New York, 1988.

[107] T. Terlaky (editor). *Interior Point Methods of Mathematical Programming*. Kluwer Academic Publishers, Boston, MA, 1996.

[108] M.J. Todd. Probabilistic models for linear programming. *Mathematics of Operations Research*, 16:671–693, 1991.

[109] M.J. Todd. A study of search directions in primal-dual interior-point methods for semidefinite programming, *Optimization Methods and Software*, 11:1–46, 1999.

[110] M.J. Todd, K.C. Toh and R.H. Tütüncü. On the Nesterov-Todd direction in semidefinite programming, *SIAM Journal on Optimization*, 8:769–796, 1998.

[111] M.J. Todd and Y. Ye. A lower bound on the number of iterations of long-step and polynomial interior-point linear programming algorithms. *Annals of Operations Research*, 62:233–252, 1996.

[112] K.C. Toh, M.J. Todd and R.H. Tütüncü. SDPT3 - A MATLAB package for semidefinite programming, version 2.1. Interior point methods. *Optimization Methods and Software*, 11/12:545–581, 1999.

[113] T. Tsuchiya. A polynomial primal-dual path-following algorithm for second-order cone programming. Technical Report 649, The Institute of Statistical Mathematics, Tokyo, December, 1997.

[114] T. Tsuchiya. A convergence analysis of the scaling-invariant primal-dual path-following algorithms for second-order cone programming. Interior point methods. *Optimization Methods and Software*, 11/12:141–182, 1999.

[115] A.W. Tucker. Dual systems of homogeneous linear relations, in: H.W. Kuhn and A.W. Tucker (editors), *Linear Inequalities and Related Systems, Annals of Mathematical Studies*, 38:3–18, Princeton University Press, Princeton, NJ, 1956.

[116] L. Vandenberghe and S. Boyd. *SP: Software for semidefinite programming*. Information System Laboratory, Electrical Engineering Department, Stanford University, CA, 1994.

[117] L. Vandenberghe and S. Boyd. A primal-dual potential reduction method for problems involving matrix inequalities. *Mathematical Programming*, 69:205–236, 1995.

[118] L. Vandenberghe and S. Boyd. Semidefinite Programming. *SIAM Review*, 38:49–95, 1996.

[119] R.J. Vanderbei. LOQO user's manual – version 3.10. Interior point methods. *Optimization Methods and Software*, 11/12:485–514, 1999.

[120] H. Wolkowicz, R. Saigal and L. Vandenberghe. *Handbook of Semidefinite Programming (Theory, Algorithms and Applications)*. Kluwer Academic Publishers, Boston, MA, 2000.

[121] S.J. Wright. A path-following infeasible-interior-point algorithm for linear complementarity problems. *Optimization Methods and Software*, 2:265–295, 1993.

[122] S.J. Wright. *Primal-Dual Interior-Point Methods*. SIAM, Philadelphia, PA, 1997.

[123] X. Xu, P.-F. Hung and Y. Ye. A simplification of the homogeneous and self-dual linear programming algorithm and its implementation. *Annals of Operations Research*, 62:151–172, 1996.

[124] Y. Ye. On the finite convergence of interior-point algorithms for linear programming. *Mathematical Programming*, 57:325–335, 1992.

[125] Y. Ye. Toward probabilistic analysis of interior-point algorithms for linear programming. *Mathematics of Operations Research*, 19:38–52, 1994.

[126] Y. Ye. On homogeneous and self-dual algorithms for LCP. *Mathematical Programming*, 76:211–222, 1997.

[127] Y. Ye. *Interior Point Algorithms, Theory and Analysis*. John Wiley & Sons, Chichester, 1997.

[128] Y. Ye, M. Todd and S. Mizuno. An $\mathcal{O}(\sqrt{n}L)$-iteration homogeneous and self-dual linear programming algorithm. *Mathematics of Operations Research*, 19:53–67, 1994.

[129] A. Yoshise. Complementarity problems, in: T. Terlaky (editor), *Interior Point Methods of Mathematical Programming*, pp. 189–252. Kluwer Academic Publishers, Dordrecht, 1996.

[130] L.L. Zhang and Y. Zhang. On polynomiality of the Mehrotra-type predictor-corrector interior-point algorithms. *Mathematical Programming*, 68:303–318, 1995.

[131] Y. Zhang. User's guide to LIPSOL: linear-programming interior-point solvers V0.4. Interior point methods. *Optimization Methods and Software*, 11/12:385–396, 1999.

[132] Y. Zhang, On extending some primal-dual algorithms from linear programming to semidefinite programming, *SIAM Journal on Optimization*, 8:365–386, 1998.

[133] G.Y. Zhao. Interior point algorithms for linear complementarity problems based on large neighborhoods of the central path. *SIAM Journal on Optimization*, 8:397–413, 1998.

[134] Y.B. Zhao and G. Isac. Properties of a multivalued mapping associated with some nonmonotone complementarity problems. *SIAM Journal on Control and Optimization*, 30:571–593, 2000.

[135] Y.B. Zhao and D. Li. Strict feasibility condition in nonlinear complementarity problems. *J. of Optimization Theory and Applications*, 107:643–666, 2000.

[136] J. Zhu. A path following algorithm for a class of convex programming problems. *Zeitschrift für Operations Research*, 36:359–377, 1992.

Index